Michaela Pelican, Karim Zafer, Michael Bollig (eds.)
Decolonising the Future Academy in Africa and Beyond

Postcolonial Studies | 54

Editorial

Postcolonial research has brought critical perspectives on colonialism in history and its heritage nowadays into public and scholarly focus. Similarly, postcolonial theorists have shown how deeply European scholarship and education are intertwined with the history and present of colonialism. For quite some time now, this postcolonial critique has triggered important public debates on the colonial past and how it should be remembered. The series **Postcolonial Studies** offers an editorial platform to continue these important discussions in an interdisciplinary framework.

Michaela Pelican, born 1972, works as a professor of social and cultural anthropology at the University of Cologne. She is the rector's representative for international affairs and the speaker of the Global South Studies Center Cologne (GSSC). She did her doctorate at the Martin-Luther University Halle-Wittenberg and was a junior researcher at the Max Planck Institute for Social Anthropology in Halle/Saale. Before coming to Cologne in 2012, she was a senior lecturer and researcher at the University of Zurich and a visiting fellow at the University of Kyoto. Her research focuses on migration, inequality, conflict, indigeneity and research methodology with a regional focus on West Africa, the Gulf States and China.
Karim Zafer, born 1981, works as a project manager and is responsible for the coordination of academic collaboration with the MENA-Region at the Divison for International Affairs at the University of Cologne. He is also a lecturer at the Department of Social and Cultural Anthropology at the University of Cologne, from which he received his PhD in 2024 and Master's degree in 2016. Before migrating to Germany in 2013, he studied and worked in the fields of tourism and Egyptology in Egypt. His research interests include (forced) migration, youth, decolonial research methodology, and the MENA region.
Michael Bollig, born 1961, works as a professor of social and cultural anthropology at the University of Cologne. He did his doctorate at the University of Tübingen on violence and conflict management in northern Kenya. He started at the University of Cologne in 2000 and established a focus on environmental challenges and socio-cultural dynamics in rural Africa. He has been speaker (and co-speaker) of the collaborative research centre (CRC) Arid Climate, Adaptation and Cultural Innovation in Africa (2000-2007) and the CRC Future Rural Africa (2018-ongoing). Between 2011 and 2015 he was vice rector for international affairs at the University of Cologne.

Michaela Pelican, Karim Zafer, Michael Bollig (eds.)

Decolonising the Future Academy in Africa and Beyond

Institutional Development and Collaboration

[transcript]

The Global South Studies Center (GSSC) coordinates global, interdisciplinary research at the University of Cologne focusing on currently highly relevant topics, such as human-environment relations, migration, inequality and knowledge production. https://gssc.uni-koeln.de/en/

Funding for this publication was provided by the Global South Studies Center (GSSC), the Collaborative Research Centre Future Rural Africa (CRC 228), and the University of Cologne. During the phase of finalising this volume, Michael Bollig benefitted additionally from the ERC Advanced Grant "Rewilding the Anthropocene", while Michaela Pelican received additional support from the Volkswagen Grant 9699-2 "Communication during and after COVID-19".

Bibliographic information published by the Deutsche Nationalbibliothek
The Deutsche Nationalbibliothek lists this publication in the Deutsche Nationalbibliografie; detailed bibliographic data are available in the Internet at https://dnb.dnb.de

transcript Verlag | Hermannstraße 26 | D-33602 Bielefeld | live@transcript-verlag.de

Cover design: Maria Arndt
Cover illustration: "La revanche, ou les Français au Missouri", 1829 Paris, France © The Trustees of the British Museum, https://www.britishmuseum.org/collection/object/P_2006-U-209
Printing: Elanders Waiblingen GmbH, Waiblingen
Copy-editing: Kim Crowder / Proofread: Sarah Mund
https://doi.org/10.14361/9783839475966
Print-ISBN: 978-3-8376-7596-2
PDF-ISBN: 978-3-8394-7596-6
ISSN of series: 2703-1233
eISSN of series: 2703-1241

Printed on permanent acid-free text paper.

Contents

Part 2:
Decolonising the Academy – Practical Examples

Acknowledgements

This volume originated from a series of three workshops and panels in 2019 (at the University of Cologne, the European Conference of African Studies in Edinburgh, and the conference of the German Anthropological Association in Konstanz) that brought together anthropologists and Africanist scholars from universities in Africa and Germany. They jointly discussed and reflected upon the relevance, challenges and potentials of social and cultural anthropology as a discipline at African universities, teaching and research collaborations within and beyond the continent, and the production of actionable knowledge and job opportunities. The workshops and panels' primary aim was to share experiences and visions of how to decolonise the academy and to learn from each other's practices and collaborations. The participants included: Munzoul Assal, Heike Becker, David Bogopa, Michael Bollig, Pierre Boizette, Rosabelle Boswell, Daniele Cantini, Yntiso Deko Gebre, Ayodele Jegede, Erik Mutisya Kioko, Thomas Kirsch, Patience Mutopo, Romie Nghitevelekwa, Isaac K. Nyamongo, Sung-Joon Park, Michaela Pelican, Ciraj Rassool, Claudia Rauhut, Nikolaus Schareika, Kira Schmidt, Antoine Socpa, Cordula Weißköppel, Ulrike Wesch, Karim Zafer. While not all participants contributed to this volume, the chapters certainly profited from in-depth discussions throughout the workshops.

It took a while to complete the volume. After the workshops and panels in 2019, most authors were confronted with the consequences of the Covid-19 pandemic. During 2023, we succeeded in bringing all contributions together, commented upon them as an editorial team and sampled the final versions of the papers. Taking the volume forward to its final, publishable state, a number of people assisted. Carola Jacobs and Jonna Bienert helped organise the workshops and panels early on. Kim Crowder copy-edited all the chapters. Through the latter stages of the publication Sarah Mund, a researcher at the University of Cologne, accompanied the becoming and finalisation of the volume.

Painstakingly, Sarah worked on the proofread and references and facilitated the final layout of the volume. She was assisted by Juliana Hagmann when controlling the bibliography for completeness and style.

The workshops and the publication of the volume were made possible through grants from the Global South Studies Center Cologne (GSSC), the Collaborative Research Centre Future Rural Africa (CRC 228, funded by the German Research Council) and the University of Cologne. During the phase of finalising this volume, Michael Bollig benefitted additionally from the ERC Advanced Grant *Rewilding the Anthropocene*, while Michaela Pelican received additional support from the Volkswagen Grant 9699–2 Communication *During and After COVID-19*. The German Association of Social and Cultural Anthropology assisted the editors of the volume with an additional grant to run one of the workshops leading to this publication.

This edited volume is the result of these ongoing conversations. In addition to the authors in this volume, we thank all the colleagues who participated in the workshops and panels and enriched our discussions with their insightful contributions.

The editorial team: Michaela Pelican, Karim Zafer, Michael Bollig

Introduction: Towards Decolonising the Future Academy in Africa and Beyond

Karim Zafer, Michaela Pelican and Michael Bollig

What does it mean to decolonise the academy in Africa? Is this a project of the future? Or are there forerunners in the history of anthropology at African universities? This book brings together authors from different parts of Africa and Germany who participated in a series of workshops and panels dedicated to these questions. Their contributions show that there are no easy answers. There have been different trajectories of anthropology as a discipline and of decolonising the academy across the continent. Similarly, the authors' assessment of the future academy diverges: While some are hopeful and take inspiration from earlier experiences of disciplinary and methodological developments, others remain critical and call for more radical attempts at decolonising the academy, not only in Africa but also in Germany.

A major theme running through the book is institutional and disciplinary developments at African universities, with our contributors focusing mostly on the trajectories of anthropology in Cameroon, Chad, Egypt, South Africa and Sudan. We argue that these institutional developments – mostly fuelled by political-economic transformations – have engendered early pathways of decolonising the academy on which current initiatives can draw. A second and key theme of the book is collaboration between African and German scholars, with examples in the fields of research, teaching, institutional development and professional networks. We believe such North-South collaborative efforts signal the way forward in decolonising the academy both in Africa and Europe.

Since the #RhodesMustFall (#RMF) movement, which ignited at the University of Cape Town in 2015 and then gained momentum across South Africa and beyond, African scholars have been vital voices in current debates on decoloniality and on decolonising the academy (Becker in this volume). African scholars, such as Rosabell Boswell (Boswell and Nyamnjoh 2018), Achille Mbembe (2021), Sabelo Ndlovu-Gatsheni (2018, 2020), Francis Nyamnjoh (2016),

Ciraj Rassool (Rassool et al. 2022) and many others, have tackled the question of how to decolonise the academy from various angles, including on a conceptual level, in research, in the classroom, in university administration, in the museum, or as public intellectuals. They all emphasise their building on the work of earlier well-known African and Afro-Caribbean scholars, statesmen and writers, such as Franz Fanon, Jomo Kenyatta, Ngũgĩ wa Thiong'o, and many more. In this volume, we ask what is the role of anthropology in these ongoing debates and processes of decolonising the academy in Africa? What can we learn from the historical trajectories of anthropology as a discipline in different parts of the continent? Are there practical examples from which we can take inspiration for the future academy in Africa?

To address these and other questions, the editors of this volume organised a series of three workshops and panels in 2019 (at the University of Cologne, the European Conference of African Studies, and the conference of the German Anthropological Association) that brought together anthropologists and Africanist scholars from universities in Africa and Germany (Jacobs and Pelican 2019). They jointly discussed the relevance, challenges and potentials of anthropology as a discipline at African universities, teaching and research collaborations within and beyond the continent, and the production of actionable knowledge and job opportunities. The workshop and panels' primary aim was to share experiences and visions of how to decolonise the academy and to learn from each other's practices and collaborations. The participants included: Munzoul Assal, Heike Becker, David Bogopa, Pierre Boizette, Michael Bollig, Rosabelle Boswell, Daniele Cantini, Yntiso Deko Gebre, Ayodele Jegede, Erik Mutisya Kioko, Thomas Kirsch, Patience Mutopo, Romie Nghitevelekwa, Isaac K. Nyamongo, Sung-Joon Park, Michaela Pelican, Ciraj Rassool, Claudia Rauhut, Nikolaus Schareika, Kira Schmidt, Antoine Socpa, Cordula Weißköppel, Ulrike Wesch and Karim Zafer. This edited volume is the result of these ongoing conversations.

The volume is structured in two parts: the first assembles contributions focusing on the trajectories of anthropology as a discipline at universities in different parts of the African continent; the second tackles the question of decolonising anthropology from methodological and practical perspectives.

Historical trajectories of anthropology at African universities

In the social sciences and humanities – and maybe to a lesser degree in the natural sciences – the dominance of the Eurocentric epistemic model has been challenged by different actors in Africa and other parts of the world (e.g. Bhambra 2014; Falola 2023; Gu 2023; Maldonaldo Torres 2016; Mignolo and Walsh 2018; Ndlovu-Gatsheni 2018; Ngũgĩ wa Thiong'o 1986; Nyamnjoh 2016; Parashar and Schulz 2021; Quijano 2000; Tangwa 1992). Students and academics have decried the colonial legacy of African universities. They have demanded a break with the past so as to develop curricula that are meaningful to the African context, and to reach out for knowledge production beyond the academy.

Generations of anthropologists have been faced with the following questions: What makes anthropology different from sociology, for example? What methods and theories do anthropologists employ that other social scientists do not habitually make use of? What social, ecological, or political dilemmas does anthropology address specifically? And how do the discipline's methods and theories contribute to an understanding of these challenges? How can graduates from anthropology Bachelor's, Honours and Master's degree courses be suitably employed, and where are future job markets? How can academics at African universities fruitfully reconcile the quest for basic research with the demand for short-term consultancy work?

While these and similar questions unite anthropologists across the continent and beyond, the discipline's trajectories have differed profoundly between various African countries. In many regions, anthropology has only recently been instituted as an academic discipline and has had to negotiate its status amidst other social sciences. In other contexts, though, anthropology has a pedigree at African universities that goes back to the 1920s (e.g. South Africa), 1930s (e.g. Egypt), 1950s (e.g. Sudan) and 1960s (e.g. Cameroon, Kenya). The discipline's relevance for addressing questions of social well-being, societal transformation or the valorisation of local and indigenous knowledge has been reflected upon profusely in African academia over many decades as well as more recently (e.g. Nkwi 2015; 2021; Ntarangwi et al. 2006; Nyamnjoh 2016).

Beyond these practical questions pertaining to anthropology's place in academia, the discipline has been faced with generalising allegations that anthropologists of the past directly contributed to colonial rule and that the discipline has never emancipated itself from the hegemony of Northern institutions of knowledge production (Asad 1979; Mafeje 1998; Nyamnjoh 2012a, 2012b). For reasons that demand further exploration, these allegations have been formulated against anthropology more vociferously than against other disciplines, for example the agricultural sciences (Beinart et al. 2009; Bonneuil 2000) or the engineering sciences (Eichhorn 2019; Muller 2018). Diane Lewis (1973) and Talal Asad (1979) denounced anthropology's leaning towards colonial administrations and its effort to become a useful science within the colonial context. Other critics depicted how anthropologists and anthropological accounts were shaped by colonial contexts, exploring patronage and funding by colonial administrations (Moore 1994; Ntarangwi et al. 2006). These authors not only highlighted the impact of colonial rule and colonial ideologies on the discipline, but also traced the influence of anthropological teaching on colonial administrators. In his presentation at one of the workshops leading up to this volume, Kenyan anthropologist Isaac Nyamongo highlighted the extent to which anthropological knowledge became essential for the training of future colonial administrators (see also Nyamongo 2007). As a consequence of this debate, in the first decades of postcolonial academia, many anthropologically educated scholars camouflaged themselves in departments of social sciences to navigate an academic environment hostile to anthropology. Some scholars, such as Mwenda Ntarangwi and colleagues (2006), argue against the oversimplified assumption of anthropology's collusion with colonial agendas. They highlight the diverse perspectives within the discipline during the colonial era, ranging from complicity to active critique and resistance against colonial ideologies.

Indeed, anthropology did not feature importantly at African universities during the first decades of post-independence (Assal 2024 for Sudan; Gebre 2024 for Ethiopia; Nyamongo 2007 for Kenya; Socpa and Nkwi in this volume for Cameroon). It was the 1980s that marked a pivotal period in African development as economic crises prompted a re-evaluation of existing paradigms, ushering in a search for alternative models. This shift created an opportune moment for anthropology to assert its relevance, with African anthropologists increasingly sought after as consultants by governments and development agencies (Nkwi 2015, 2021; Ntarangwi et al. 2006). As the focus turned towards restructuring African economies in the 1990s, anthropology found itself at the forefront of shaping developmental agendas.

In their contribution, **Dynamics of Anthropology Teaching and Practice in Cameroonian Universities (1962–2023)**, Antoine Socpa and Paul Nkwi describe how the academic discipline of anthropology has evolved in Cameroon since the early 1960s when it was first taught as an integral part of sociology and subsequently instituted as an independent discipline after nation-wide university reforms in the early 1990s. With the expansion of the Cameroonian university network, the discipline of anthropology became firmly established in several universities. With an initial enrolment of about 50 students during the 1993–1994 academic year, the Department of Anthropology at the University of Yaoundé I – the oldest and largest in the country – currently has about 750 students regularly enrolled in the Bachelor's, Master's and Doctorate cycles. These figures alone underline that anthropology is highly attractive to Cameroonian students. As the authors show, the anthropology taught and practiced in Cameroon has fostered a distinct decolonial approach that foregrounds the practical relevance of anthropological knowledge for meaningful development and social change. With the creation of new and applied subdisciplinary specialisations, such as development anthropology and medical anthropology, the discipline responded to broader political and societal needs and has become an integral part of social science training in public and private universities and professional schools. Anthropology training curricula in Cameroonian universities emphasise applied issues. Practical training aims to provide professional skills that transform students into cultural brokers for development and consultants for organisations seeking their expertise.

A strong emphasis on applied perspectives of anthropology was also portrayed by Isaac Nyamongo for Kenya, Monzoul Assal for the Sudan and Yntiso Deko Gebre for Ethiopia in their oral presentations to the 2019 workshop leading to this volume (see also Assal 2024; Gebre 2024; Nyamongo 2007). Notably, Nyamongo pinpointed that in the Kenyan context, there is acute pressure on anthropology to produce actionable knowledge. A similar assessment is shared by Remadji Hoinathy and colleagues for Chad (in this volume). Recent calls for decolonising the academy have highlighted not only the relevance of epistemic freedom, but also the vital role and expertise of African scholars in both questioning and defining the meaning and future of development in Africa (Ndlovu-Gatsheni 2020; Olukoshi 2006). Ntarangwi et al. (2006) certainly have a valid point when arguing that consultancies run the risk of confining intellectual production to routine reports, hence sacrificing scholarly creativity to survival necessities. However, the widespread acknowledgement of anthropological knowledge as a meaningful contribution to development projects across the continent has brought about broad academic and political acceptance, opened up new venues for job-seeking young anthropologists and, in many countries, led to the inauguration of new anthropology departments (Hoinathy et al. in this volume; Socpa and Nkwi in this volume; Nyamongo 2007).

Regarding Egyptian anthropology, Daniele Cantini critically reflects upon the problematic aspects of undue government impact on the contents of anthropological research and anthropological teaching. Cantini's contribution, **Notes on the Institutional Development of Anthropology in Egypt and toward its Decolonisation** depicts the development of the discipline in Egypt and its contemporary attempts to adapt to quests for a national Egyptian identity in the context of contradictory trends towards globalisation and Islamisation, and the socio-economic transformation and readjustments of academic institutions after the 2011 revolution and its aftermath. As in South Africa (Becker in this volume), anthropology entered the academic syllabus in Egypt early on in the 20th century. Unlike South Africa, however, the discipline did not become well established at Egyptian universities; neither during colonial times, nor through the times of socialist Nasserite nationalism, nor in the post-Nasser period. Prominent Egyptian anthropologists obtained their training abroad, particularly in Europe, and subsequently worked in Egypt in various roles. The country's first anthropology department was established at the University of Alexandria in the 1970s. The department produced a small number of graduates and PhD holders who were instrumental in setting up anthropology departments at a couple of other Egyptian universities (Cairo University, Ain Shams University). The staff were joined by further PhD holders returning from various European universities. However, anthropology did not flourish. In fact, it was further marginalised, and over the past two decades, the number of graduates emerging from anthropology departments has reduced significantly. Cantini surmises that the development of anthropology in Egypt has suffered from a number of contradictory demands on the discipline: following the events of 2011, how could anthropology meaningfully contribute to the social and cultural development of a country which has been shaped by so many polarising debates about possible futures? If there was an anthropological contribution to this end, was this reflexive and critical, or rather, a social-engineering endeavour?

While nationalist agendas can play a crucial role in shaping the discipline, international connections – often rooted in colonial legacies – are equally important. These dynamics yield both positive and negative outcomes for the discipline. On the one hand, African anthropology has flourished through interactions with British, French, American and other European anthropologies. These exchanges encompass various forms, such as scholarships, sabbaticals, workshops, conferences and joint research projects. Such engagements have shielded African anthropology from provincialism, enriching it with diverse perspectives and methodologies (Ntarangwi et al. 2006). On the other hand, this international focus has come at the expense of fostering similar connections among African anthropologists, even within the confines of a single country. It is striking to note that communication and collaboration across national boundaries among African anthropologists are often less frequent than those with colleagues in Europe and North America. This disconnect is particularly pronounced between Anglophone and Francophone African academics, where language acts as a significant barrier (ibid.). Several of the contributions to this volume attest to African universities' vibrant national and international networks and their vital role in promoting disciplinary development (Becker for South Africa; Hoinathy et al. for Chad; Pelican and Ngeh and Socpa and Nkwi for Cameroon). Yet besides such university-to-university collaborations, the establishment of professional organisations and networks on the continent, such as the Council for the Development of Social Science Research in Africa (CODESRIA), have been key steps toward strengthening the visibility and epistemic freedom of African scholarship (Olukoshi and Nyamnjoh 2006; Nkwi 2018).

In their contribution, **Thoughts on "Coloniality" and Africanity: Scholarship in African Universities and African Studies in Germany,** Abimbola Adesoji and Hans Peter Hahn critically reflect upon the foundations of Northern social science and humanities thought within African academia. They argue that framing research on Africa and Africans and the rise of scientific disciplines elaborating knowledge on the continent and its inhabitants was invariably embedded in epistemological traditions of Europe established during the 16th to 18th centuries. In the 19th century, imperialism and colonial expansion used these frameworks as a justification for political subjugation and exploitation. A cornerstone of decolonial efforts is the exposure of the historical foundations and gross power imbalances of disciplinary thought and practices. Epistemological violence and flagrant disregard for alternative non-European perspectives were essential building blocks during the laying of disciplinary foundations. Adesoji and Hahn delineate what the efforts that undermine such academic foundations framed by coloniality might look like. Systematic consultation with African humanities scholars – for example, in the fields of history, philosophy and social and cultural anthropology – and the integration of alternative theories and scientific practices into the canon of disciplines are essential for advancing the emerging decolonial project. Adesoji and Hahn single out the German African Studies Association (VAD) as a case study and establish both its historical foundations and its original framing through a coloniality mindset. But did emancipatory efforts made by the organisation in the second half of the 20th century help to replace such earlier frames? Adesoji and Hahn are sceptical and show that despite considerable effort to engage with decolonial perspectives, the organisation retains some of its former frames. They argue that it is the task of the contemporary generation of social scientists to further the project of decoloniality and to give it a face; that is, to translate a critical and reflexive mindset into meaningful institutional solutions that can truly transform the epistemic foundations of the social sciences and humanities.

Decolonising the academy – practical examples

Part two of the edited volume discusses strategies concerned with addressing the challenge of decolonising anthropology in practical terms. Three approaches are taken up: changing the curriculum, international collaboration at par, and the establishment of a decolonised "native" anthropology.

African anthropologists' development of meaningful intellectual agendas, working practices and international collaborations have progressively contributed to attracting increasing numbers of students (Ntarangwi et al. 2006; Nyamnjoh 2016). Despite these obvious successes, forging an identity for anthropology in Africa remains a complex task since many of its proponents have been trained in traditions of scholarship at institutions in the Global North. As in other academic disciplines (Chaya and Chika 2018; Molla et al. 2016), many African early career scholars earn their doctorate at universities in Europe and North America, while the number of PhD graduates from Indian and Chinese universities has been increasing in recent decades. Until now, PhD studies financed at universities of the Global North have provided for more financial stability and more research funding than universities in Africa can habitually supply. At the same time, African anthropologists seek to escape the academic treadmill of engaging with theoretical debates mainly taking place in Europe and North America, and actively contribute to decolonising the discipline and its curricula (Becker in this volume; Jegede 2015).

Heike Becker's contribution, **"Because Rhodes Fell": Historical development, institutional contexts, and the challenges of decolonisation of South African anthropology,** opens with reflections on the state of anthropology in South Africa and the discipline's attempt to consider aspirations towards decolonisation in its teaching and research practices during the 2010s. While the quest for decolonisation currently prevails in discourses at and around many anthropology departments in South Africa, the target of change is a moving one: What exactly needs to be decolonised? How is this to be done best and who are the actors of decolonisation? There is no shortcut to decolonisation, no blueprint on how to go about it, and no single formula on what to address first. Becker describes in great detail what single anthropology departments do to experiment in order to approach the needs and wants of decolonisation adequately. It is especially this part of Becker's contribution that is of broader interest. The first part of the chapter provides a concise overview of the discipline's institutional development spanning the colonial, apartheid and early post-apartheid eras. As Becker rightly points out, anthropology as a discipline has a long history in South Africa dating back to the 1920s. It has produced many influential and critical thinkers, including White and Black South African anthropologists, who in their work have been driven by the country's particular political-economic context and its radical transformation. Becker provides an insightful account of early Black anthropologists, many of whom were forced to leave the country in the course of their careers. Black anthropologists gained prominence again in the post-apartheid era and, importantly, in the 21st century with student protests and the #RhodesMustFall (#RMF) movement which explicitly urged the decolonisation of South African academia and society. The second part of Becker's contribution sketches a number of approaches to instil decolonisation into the syllabus and the staffing structure of departments. The challenge arises from the fact that there is no single meaning of decolonisation. Consequently, numerous critiques and strategies have emerged, including calls for curriculum reform, reassessment of research subjects and objectives, and embracing new forms of knowledge and anthropological writing genres. Giving detailed accounts of different departments and colleagues' initiatives, Becker outlines approaches towards readjustment of the discipline's curriculum and pedagogical practice that enable it to address the broader challenges of social transformation. She ends with the conscious remark that decolonisation is a truly transnational endeavour, thus echoing the call of Adesoji and Hahn (in this volume) to decolonise the academy not only in Africa but also in Germany.

With regard to decolonising the curricula, we can also draw on the experiences of different national traditions, for example Cameroon and Kenya, that long have gone their own way of "Africanising" their anthropology curricula, particularly in regard to their applied strands of development and medical anthropology. Yet, as Socpa and Nkwi (in this volume) emphasise, it is key to retain also "pure"/classical anthropology and an international corpus of key readings as part of the general training in order to produce high quality research, which is also the basis for good applied research.

In his contribution, **(Re)Building Epistemology or (Re)Shaping Societal Outlook: A critique of the Sudan's Islamisation of knowledge paradigm,** Bakheit Mohammed Nur deals with another more radical approach to decolonisation. In Sudan, not only were the curricula of some social science disciplines transformed, but, through a programme that aimed at the "Islamisation of knowledge", decolonisation envisioned a radical rejection of a scientific paradigm that was framed as "Euro-American" in favour of Islamic epistemologies. This politically motivated programme aspired to tie academic excellence to an Islamisation of all scientific disciplines. Islamic sciences, Islamic educational institutions, publication houses and academic journals were thus intended to contribute to well-being, political unity and sustainability. On the basis of fieldwork in Khartoum State, which is the home of three major Islamic universities – Omdurman Islamic University, the University of Holy Qur'an and Islamic Sciences, and the International University of Africa – Nur reflects upon the dilemmas of a wholesale decolonisation via Islamisation. Will the formulation of Islamic concepts for all disciplinary specialisations and a thorough scrutiny of Euro-American concepts suffice? Or is there a need to reorganise disciplines altogether and opt for a new division of labour among the social sciences and, in the end, establish new disciplines based on Islamic epistemologies? Nur shows that the "Islamisation of knowledge" project is politically motivated and that its contents and procedures are also contested amongst Islamic researchers. Some argue that the basic scientific structures are fundamentally shaped by the thinking of enlightenment and modernisation. To cut oneself off from this fundamental disciplinary organisation of academia is tantamount to self-inflicted marginalisation. Others argue that the "Islamisation of knowledge" project favours one version of Islam over a diversity of Islamic beliefs and approaches to knowledge and, hence, disregards the breadth and multitude of Islamic civilisations.

While the transformation of teaching syllabi or of entire disciplines are possible trajectories for decolonising the future academy, other approaches tackle the question of how to decolonise research methods and collaboration (Guma et al. 2024; Mogstad and Lee-Shan 2018). For example, Francis Nyamnjoh (2012b) advocates for a paradigm shift in anthropology's approach to studying Africa, emphasising the importance of embracing and reflecting the continent's creative diversity. He argues that anthropologists studying Africa should incorporate this diversity into every aspect of their research projects, from conceptualisation to implementation, and in their collaboration with "native" and "at-home" anthropologists across disciplines. Using the metaphor of the three blind men and the elephant, he also rejects the idea of replacing one perspective with another, such as that of the European anthropologists (blind men), with that of the "natives" (the elephant). Instead, he advocates for facilitating conversations that embrace diverse viewpoints and acknowledge the limitations inherent in each perspective. This requires a critical interrogation of anthropologists' assumptions that our ways of knowing are superior to the ways of knowing of fellow academics, or of those we study.

Central to Nyamnjoh's proposal is the concept of co-production, which goes beyond conventional professional collaboration to prioritise teamwork and inclusivity. He calls for multi- and transdisciplinary endeavours that actively involve the communities being studied in shaping the research process. This approach aims to create a more equitable platform for dialogue among multiple perspectives, moving away from the hierarchical dynamics of earlier anthropological research.

The contribution of Michaela Pelican and Jonathan Ngeh, **Towards Joint Production of Knowledge in the Third Space – lessons learned,** critically reflects on the challenges and benefits of collaboration as a possible pathway toward decolonising anthropological research. How can students and scientists from different contexts work together despite power asymmetries? Pelican and Ngeh discuss efforts towards joint and cooperative knowledge production in a student research programme that was co-organised by the Cameroonian Universities of Yaoundé, Dschang and Bamenda, and the German University of Cologne. Six Cameroonian students and six German students accompanied by five established academics set out to research the future-oriented aspirations of urban youth in Yaoundé, Cameroon's capital. The authors take Homi Bhabha's (2004) concept of Third Space as a lead concept. Do cultures mix and is hybridisation possible as "a double process of decontextualization and recontextualization, first reaching out to appropriate something alien and then domesticating it" (Burke and Hsia 2007: 10)? Such translation work is intended to open up options for something new. Pelican and Ngeh share this optimistic assessment of hybridity and engage the reader in a detailed, engaged and well reflected ethnography of cooperation. Both authors reflect upon their (different) perspectives and engagements with the collaborative project. Ngeh joined the project as a non-tenured lecturer at a late stage, while Pelican was a full professor as well as being in charge of the programme's budget which came from the German University entirely. Their contribution is innovative as it does not fuse these different perspectives in one text, but presents two different accounts of the same process. In this way, they each reflect separately upon the preparation process, the fieldwork period and the final data analysis before coming to a joint conclusion. Pelican and Ngeh show that such collaboration is possible if asymmetries of power are made visible and are reflected upon. Academic cooperation across borders is certainly one important instrument of decolonising the social sciences in general and anthropology in particular. Hence, the auto-ethnographic approach adopted here makes problematic constellations visible and does not deny them in a superficial discourse on collaboration. Power asymmetries, of course, do not collapse if we only want them to disappear. Pelican and Ngeh show that a consciousness for, and a critique of, such power imbalances is essential; but beyond that, only intense social work – communication and practical collaboration – can lead towards joint knowledge production in a social space.

In a similar vein the contribution by Remadji Hoinathy, Djimet Deli and Andrea Behrends, **Doing Anthropology at Home, in Chad,** depicts how the collaboration between researchers from Chad and various academic institutions in Europe gave rise to the development of a national anthropology in Chad. The authors convincingly argue that this collaborative effort enabled Chadian anthropology to develop a decolonial perspective and to test-run innovative methods of engaged empirical research. The decolonising impetus is based on the well reflected argument that, in the Chadian context, anthropology must first of all contribute to societal ends, to more justice, to more participation and to more recognition. Chad's young and emergent anthropology must fulfil these aims in the face of governmental mistrust and neglect. Indeed, local political conditions characterised by authoritarianism, clientelism and dependency are described as a major constraint for the development of a national anthropological tradition. In other words, Chadian anthropologists have to cope with an adverse political environment and, at the same time, justify their way of doing anthropology vis á vis international partners and donors who prioritise basic research and are often averse to application-focussed research. Hoinathy and colleagues describe how Chadian anthropology developed on the institutional level as a separate discipline taught at the national university and at the same time successfully established the non-university anthropology centre CRASH (*Centre de Recherches en Anthropologie et Sciences Humaines*). The great number and immense diversity of CRASH's activities give evidence of the country's need for anthropological knowledge and ethnographic methodology in numerous societally highly relevant fields, ranging from the social-ecological consequences of Chad's oil boom to the proliferation of violent conflicts. In many such fields a devoted, application-focussed anthropology substantially adds to the understanding of local concerns and local adaptive strategies.

A third approach towards decolonising methodology outlined in this volume is the establishment of a decolonised "native" anthropology. Is doing "anthropology at home" in Africa recognised, accepted and independent? Ntarangwi et al. (2006) believe that African anthropologies currently being carried out "at home" have a real value and are crucial for the future of the discipline. Yet, they argue that amidst the international exchanges that take place, African anthropology finds itself at a crossroads, torn between asserting its own identity and building upon the established traditions of scholarship present in Global North institutions where its practitioners have been trained.

Faced with similar challenges, Souleymane Diallo and Karim Zafer problematise the age-old insider-outsider debate that anthropologists have contemplated for a couple of decades. Adopting a decolonial lens in their contribution, **Decolonising the "Native", Insider and Outsider Categories in Anthropology,** Diallo and Zafer reflect upon their respective fieldwork experiences. Diallo worked with different Tuareg communities in and around refugee camps in Mali, whereas Zafer worked in Germany with male unaccompanied refugee youths from Arabic speaking countries. In both instances, they show that the insider-outsider dichotomy is invalid and misleading. They first look at language competence. Too often, an excellent command of the local language is attributed to anthropologists who work within their own country. Yet, Diallo explains that his Malian identity did not guarantee any competence in Tamashek, the Tuareg language, nor did the Tuareg he interviewed necessarily request that interviews be conducted in their own language. Many were happy and competent enough to converse in French while others preferred to speak in Bamankan, the national language. Diallo and Zafer then iterate other aspects of researcher-researched relations and point to a complexity that the simplistic insider-outsider dichotomy cannot address. Although Diallo and Zafer perhaps had some "insider-advantages" when discussing things with their interlocutors, their alleged position as insiders also brought about some challenges. Zafer, for example, reports that the fact that his respondents automatically linked him with an Arabic mainstream culture made it nearly impossible to discuss things either with young women in general, or with the female partners of his interlocutors in particular. Diallo rapports that low-caste Tuareg actively tried to prevent him from any engagement with upper-caste Tuareg. The challenge both authors see is that their anthropological fieldwork environment sometimes forces them back to their "insiderhood". They are deemed to be insider experts for their societies. By conclusion this, however, also means that they are fixated on "their societies" for their research. The chapter's authors argue that "this association and ascription has been the easiest and maybe the only open way until now to enter the academic field in the Global North and to prove ourselves as trustworthy anthropologists and scholars."

Their contribution presents a strong argument for debunking the naïve and misleading insider-outsider concept and excluding it from the anthropological canon in order to allow multiple connections between researcher and researched. It also points to the need to decolonise and equalise research choices in the Global South and North and encourage researchers from the Global South to study topics and societies in the Global North. Other authors have argued in a similar direction, problematising the position of insider anthropologists. According to Nyamnjoh (2012b), acknowledging the possibility of "native" anthropology and actually recognising the achievements of "native" anthropologists are two distinct challenges. Even if individuals from marginalised backgrounds hold "residence permits" or "passports" within the anthropological community, they may still find themselves relegated to the margins, perceived as second-class citizens among their peers – figuring as the "outsider within". African anthropologists, in particular, face a unique dilemma as they navigate the desire for integration, interconnection, and interdependence within the global anthropological community.

Ways forward

We opened this introduction by asking whether decolonising the academy is a project of the future. Drawing on the contributions to this volume, we can confidently say that there have been many forerunners in different parts of the African continent, from which we can learn and draw inspiration. In some chapters (e.g. Adesoji and Hahn; Hoinathy et al.; Pelican and Ngeh), collaboration emerges as a common theme on possible ways forward to decolonising research praxis and knowledge production. However, collaboration is an active endeavour whose success is not guaranteed. We believe that decolonising the academy is a two-way process that is not limited to universities in Africa but is as much a duty and endeavour of universities in the Global North. We thus suggest taking the conversations initiated in these workshops, and continued in this edited volume, as a starting point and stimulus for further debates on decolonising the future academy not only in Africa but at German universities.

Bibliography

Asad, T. (1979): "Anthropology and the Colonial Encounter." In: G. Huizer/B. Mannheim (eds.), The Politics of Anthropology: From Colonialism and Sexism Toward a View from Below, Berlin, New York: De Gruyter Mouton, pp. 85–94. https://doi.org/10.1515/9783110806458.85.

Assal, M.A.M. (2024): "Sudan, Anthropology in." In: H. Callan (ed.), The International Encyclopedia of Anthropology, https://doi.org/10.1002/9781118924396.wbiea2101.

Beinart, W./Brown, K./Gilfoyle, D. (2009): "Experts and Expertise in Colonial Africa Reconsidered: Science and the Interpretation of Knowledge." In: African Affairs, 108/432, pp. 413–433. https://doi.org/10.1093/afraf/adp037.

Bhabha, H. K. (2004): The Location of Culture. (2nd ed.), London: Routledge.

Bhambra, G. K. (2014): "Postcolonial and Decolonial Dialogues." In: Postcolonial Studies 17/2, pp. 115–121. https://doi.org/10.1080/13688790.2014.966414.

Boswell, R./Nyamnjoh, F. (eds.) (2018): Postcolonial African Anthropologies, HSRC Press.

Bounneuil, Ch. (2000): Development as Experiment: Science and State Building in Late Colonial and Postcolonial Africa, 1930–1970. In: Osiris, 2nd Series 15/1, Nature and Empire: Science and the Colonial Enterprise, pp. 258–281. https://doi.org/10.1086/649330.

Burke, P./Hsia, R. P. (2007). "Introduction." In: P. Burke/ R.P. Hsia (eds.), Cultural Translation in Early Modern Europe, Cambridge University Press, pp. 1–4.. https://doi.org/10.1017/CBO9780511497193.

Chaya, H./Chika S. (2018): Research and PhD Capacities in Sub-Saharan Africa: South Africa Report. https://www.britishcouncil.org/sites/default/files/h233_02_south_africa_final_web.pdf.

Eichhorn, S. J. (2019): "How the West Was Won: A Deconstruction of Politicised Colonial Engineering." In: Political Quarterly 91, pp. 204–209. https://doi.org/10.1111/1467-923X.12773.

Falola, T. (2023): Sabelo Ndlovu-Gatsheni and African Decolonial Studies, London: Routledge. https://doi.org/10.4324/9781003449775.

Gebre, Y.D. (2024): "Ethiopia, Anthropology in." In: H. Callan (ed.), The International Encyclopedia of Anthropology, https://doi.org/10.1002/9781118924396.wbiea1985.

Gordon, R. (2024): South Africa's Dreams. Ethnologists and Apartheid in Namibia, Windhoek: University of Namibia Press.

Gordon, R./Spiegel, A. (1993): "Southern African Revisted." In: Annual Reviews in Anthropology 22, pp. 83–105. https://doi.org/10.1146/annurev.an.22.100193.000503.

Gu, M. D. (2023): "Is Decoloniality a New Turn in Postcolonialism"?" In: Culture as Text 1/1-2, pp. 5–18. https://doi.org/10.1515/cat-2023-1009

Guma, P. K./Akello, G./Ndlovu-Gatsheni, S. J. (2024): "Forum I: Decolonising Academic Cooperation." In: A. Fleschenberg/K. Kresse/R. C. Castillo (eds.), Thinking with the South: Reframing Research Collaboration amid Decolonial Imperatives and Challenges, Berlin, Boston: De Gruyter, pp. 63–94.

Jacobs, C. /Pelican, M. (2019): "Workshop der RG Afrika: Towards an Independent Anthropology at African Universities: Possibilities, Challenges and Trajectories." In: Deutsche Gesellschaft für Sozial- und Kulturanthropologie, Mitteilungen Nr. 52, September 2019, pp. 27–28.

Jegede, A. S. (2015): "From Gateway to Gatekeeper: Anthropology: Yesterday, Today and Tomorrow." In: P. N. Nkwi (ed.), The Anthropology of Africa: Challenges for the 21st Century, Langaa RPCIG, pp. 3–12. https://doi.org/10.2307/j.ctvh9vxg1.4.

Lewis, D. (1973): "Anthropology and Colonialism." In: Current Anthropology 14/5, pp. 581–602. https://doi.org/10.1086/201393.

Mafeje, A. (1998): "Anthropology and Independent Africans: Suicide or End of an Era?" In: African Sociological Review 2/1, pp. 1–43.

Maldonado-Torres, N. (2016): Outline of Ten Theses on Coloniality and Decoloniality, Frantz Fanon Foundation, https://www.fondation-frantzfanon.com/wp-content/uploads/2018/10/maldonado-torres_outline_of_ten_theses-10.23.16.pdf.

Mbembe, A. (2021): Out of the Dark Night: Essays on Decolonization, New York: Columbia University Press.

Mignolo, W. D./Walsh, C. E. (2018): On Decoloniality: Concepts, Analytics, Praxis, Durham, London: Duke University Press. https://doi.org/10.1215/9780822371779.

Mogstad H./Lee-Shan, T. (2018): "Decolonizing Anthropology: Reflections from Cambridge." In: The Cambridge Journal of Anthropology 36/2, pp. 53–72. https://doi.org/10.3167/cja.2018.360206.

Molla T/Cuthbert D/DJPFiE C (2016): "In Pursuit of the African PHD: a Critical Survey of Emergent Policy Issues in Select Sub-Saharan African Nations, Ethiopia, Ghana and South Africa." In: Policy Futures Educ 14, pp. 635–54.doi:10.1177/1478210316641.

Moore, S. (1994): Anthropology and Africa: Changing Perspectives on a Changing Scene, Charlottesville: University of Virginia Press.

Muller, M. (2018): "Decolonising Engineering in South Africa – Experience to Date and Some Emerging Challenges." In: South African Journal of Science 114/5-6, pp. 1–6. https://doi.org/10.17159/sajs.2018/a0270.

Narayan, K. (1993): "How Native Is a "Native" Anthropologist?" In: American Anthropologist 95/3, pp. 671–686. https://doi.org/10.1525/aa.1993.95.3.02a00070.

Ndlovu-Gatsheni, S. (2018): Epistemic Freedom in Africa: Deprovincialization and Decolonization, London: Routledge. https://doi.org/10.4324/9780429492204.

Ndlovu-Gatsheni, S. (2020): Decolonization, Development and Knowledge in Africa Turning Over a New Leaf, London: Routledge. https://doi.org/10.4324/9781003030423.

Nkwi, P. N. (2015). "Introduction: The Anthropology of Africa: Challenges for the 21st Century." In: P. N. Nkwi (ed.), The Anthropology of Africa: Challenges for the 21st Century, Langaa RPCIG, pp. ix–xiv. https://doi.org/10.2307/j.ctvh9vxg1.3.

Nkwi, P.N. (2018): "Pan-African Anthropological Association (PAAA)." In: H. Callan (ed.), The International Encyclopedia of Anthropology, https://doi.org/10.1002/9781118924396.wbiea1823.

Nkwi, P. N. (2021): From Handmaiden of Colonialism to Esteemed Discipline: Professor Paul Nchoji Nkwi on the Reinvention of Anthropology in Africa (In Conversation with Ivoline Kefen Budji), Langaa RPCIG, Bamenda, Cameroon.

Ntarangwi, M./Mills, D./Babiker, M. H. M. (2006): African Anthropologies: History, critique, and practice, London: Zed Books.

Ngũgĩ wa Thiong'o (1986): Decolonizing the Mind: The Politics of Language in African Literature, Pourtsmuth N. H.: Heinemann.

Nyamnjoh, F. B. (2012a): "Potted Plants in Greenhouses: A Critical Reflection on the Resilience of Colonial Education in Africa." In: Journal of Asian and

African Studies, 47/2, pp. 129–154. https://doi.org/10.1177/002190961141724 0.

Nyamnjoh, F. B. (2012b): "Blinded by Sight: Divining the Future of Anthropology in Africa." In: Africa Spectrum, 47/2/3, pp. 63–92. https://doi.org/10.11 77/000203971204702-304.

Nyamnjoh, F. B. (2016): #RhodesMustFall: Nibbling at Resilient Colonialism in South Africa, Langaa RPCIG. https://doi.org/10.2307/j.ctvmd84n8.

Nyamongo, I. K. (2007): "Teaching and Training in Anthropology in Kenya: The Past, current Trends and Future Prospects." In: African Anthropologist 14, pp. 19–42. https://doi.org/10.4314/aa.v14i1-2.57726.

Olukoshi, A. (2006): "African Scholars and African Studies." In: Development in Practice 16/6, pp. 533–544. https://doi.org/10.1080/09614520600958116.

Olukoshi, A./Nyamnjoh, F. B. (2006): "CODESRIA: 30 years of Scholarly Publishing." Africa Media Review 14/1-2, pp. 17–26. https://doi.org/10.57054/amr.v14i1-2.5196.

Parashar, S./Schulz, M. (2021): "Colonial Legacies, Postcolonial 'Selfhood' and the (Un)doing of Africa." In: Third World Quarterly 42/5, pp. 867–881. https://doi.org/10.1080/01436597.2021.1903313.

Quijano, A. (2000): "Coloniality of Power, Eurocentrism, and Latin America." In: Nepantla: Views from South 1/3, pp. 568. https://doi.org/10.1177/02685 80900015002005.

Rassool, C./Uzer, E./Hammami, F. (2022): "The Epistemic Work of Decolonization and Restitution: A Critical Conversation." In: F. Hammami/E. Uzer (eds.), Theorizing Heritage through Non-Violent Resistance, Palgrave Studies in Cultural Heritage and Conflict, Cham: Palgrave Macmillan. https://doi.org/10.1007/978-3-030-77708-1_10.

Tangwa, G.B. (1992): "Colonial Legacy and the Language Situation in Cameroon." In: Quest: Philosophical Discussions 6/2, pp. 24–43.

Part 1:
Historical Trajectories of Anthropology at African Universities

Dynamics of Anthropology Teaching and Practice in Cameroonian Universities (1962–2023)

Antoine Socpa and Paul Nchoji Nkwi

Introduction

The first African universities were established to provide skilled labour for the newly independent nation-states (De Haas 1994). This applies also to Cameroon, where between 1960 and 1980, university graduates easily found employment in the civil service. Even those with degrees in anthropology were quickly employed in the various ministries. Unemployment was not a problem, as vocational schools recruited graduates from all levels of education.

In the late 1970s and early 1980s, the civil service was saturated with a critical mass of university graduates and began to absorb graduates selectively through a competitive process. Those who failed floated on the labour markets. By 1986, the employability of university graduates had declined considerably. When President Paul Biya signed a decree allowing the civil service to recruit 1,500 university graduates, many found employment in the civil service with little or no substantive work experience. As the structural adjustment programmes of the 1980s imposed several restrictions on recruitment, the state was forced to reduce its recruitment.

In a major publication on the university reform, it is stated clearly that the 1993 university reforms aimed to produce graduates with practical skills that could generate independent employment (Republic of Cameroon 1993). The reforms brought together sister disciplines to provide a broad-based education. Major combined degrees such as History-Geography and Philosophy-Sociology were designed to train teachers for secondary schools. Some courses in anthropology were also taught as part of the new combined degree in philosophy and sociology. While philosophy became a subject for teaching in secondary schools, anthropology or sociology did not. Students who obtained a combined degree in Philosophy-Sociology went on to graduate studies in anthropology, but ended up as teachers in secondary schools.

The reforms provided only a short-lived solution to a growing unemployment problem. By the late 1980s, the educational field was saturated to the point where the state required universities to design curricula to produce graduates with self-employment skills. In addition, professionalisation efforts sought to provide the practical skills vital for self-employment.

For anthropology, the question was how the discipline could offer vital and critical survival skills to students. The collapse of modernisation theory in the 1960s had already generated a new debate on alternative responses to development issues. This debate opened a new window of opportunity for anthropology. Anthropology's ideas and micro-analytical perspective were explored as alternative paths to development. Would anthropology offer the answers? How could anthropology respond to the growing demands of development agencies?

In response to these questions, anthropology programmes at the University of Yaoundé and in other universities across the country attempted to address these concerns. The classical training in anthropology has been redesigned in light of the new requirements, that is, to train students with vital survival skills. The university, as a realm of ideas and idealism or a marketplace of ideas, has opened up to the demands of the world of actions, policies and practices (Nkwi 2006). While our colleagues in North America and Europe argued that applied anthropology was not anthropology, those of us who taught and interacted with the world of action believed that anthropology can only be meaningful and valuable if it listens to the everyday voices of communities and developers bringing about meaningful change. Academic anthropology has indeed inextricably been linked to development anthropology. Moreover, for anthropology to rehabilitate the colonial image in Africa, it must use its knowledge to address the many and diverse issues facing the social transformation of the continent.

Historical review of the development of the discipline in Cameroon

The teaching of anthropology at the University of Yaoundé dates back to the creation of this institution. Its status and existence today as an autonomous discipline within the Cameroonian university system is the result of a long struggle. It is, therefore, essential to evoke history to understand the motives and context that favoured the reforms and changes which subsequently affected the teaching of anthropology at the University of Yaoundé and the other public universities created since the 1993 reform.

Historically, anthropology in France and the rest of Europe referred to physical anthropology, while social and cultural anthropology was considered a sub-discipline of sociology or the French school of *ethnologie*. The role played by anthropology during the colonial period created a climate of suspicion detrimental to the discipline. This also applied to Cameroon where anthropologists were employed by the French and British colonial administrations. Accused of having served as a secular arm of the colonial enterprise (Talal Asad 1973), anthropology was regarded and described by African intellectuals, academics and nationalists as a colonial and even dangerous discipline that was unrelated to sociology. Conversely, sociology was the discipline perceived as being of great use in the process of African modernisation.

The struggles of Cheikh Anta Diop and other African scholars to restore the dignity of African cultures became a great inspiration to those who continued to study anthropology despite hostility. Even though great personalities such as Busia and Kenyatta, both trained anthropologists, had become political leaders in their respective countries by the 1960s, this fact did not help to rehabilitate the image of anthropology. It was generally deemed that anthropology was interested in tradition and thus backwards looking. Hence it was not thought to contribute to the development of postcolonial Africa. As a result, many Africans who had studied anthropology in North American and European universities were confronted with a university system that was hostile to the discipline when they returned home. By contrast, the label of the sociologist was more acceptable and marketable than that of the anthropologist.

At the University of Yaoundé, anthropology started modestly in the form of introductory courses on family, marriage, kinship and other related subjects between 1962 and 1980. At the time, the University of Yaoundé was the only public university in the country. The teaching programmes were designed in French and the overwhelming majority of teachers and students were from French-speaking Cameroon (Marchand 1962). In an attempt to accommodate students of English-speaking origin from the country's North-West and South-West regions, bilingual training courses were provided. The objective was to reduce the linguistic divide between French and English-speaking lecturers and students. It is for this reason that the University of Yaoundé has since been considered a bilingual university.

As from 1993–1994, after the university reform, anthropology gained importance with the enrolment of an increasing number of anthropology students and, consequently, the recruitment of university lecturers teaching anthropology. As most lecturers were trained in France, anthropology was taught following the French school of *ethnologie*. Only in the late 1970s, with the recruitment of Paul Nchoji Nkwi as assistant professor trained in Switzerland, was the curriculum enlarged to include other strands of anthropology, such as the British and US American strands of social and cultural anthropology (Keesing 1981). In the last two decades, culture as a fundamental concept has been the basis for the local rooting of anthropology, echoing Goodenough (1970), who approached culture from a cognitive perspective and defined it as a sum of representations, ideas and symbols shared by members of society.

As a result of the 1993 university reform, the teaching of anthropology was formalised with the creation of the Department of Sociology and Anthropology. In 2007, an independent Department of Anthropology was created. Nowadays, the specialisations offered to students in the ten state universities in Cameroon are cultural, medical, development, visual and political anthropology. The remainder of this article examines the events that have shaped the increasing visibility of anthropology in Cameroon through the various university reforms (1993, 2007 and 2023).

The 1993 university reform

From 1990 onwards, the number of students at the University of Yaoundé increased significantly. Built at the end of the 1962 to accommodate 10,000 students, the student population in the 1991/92 academic year was estimated at more than 50,000. Faced with this constant evolution in the number of students, since 1982, following numerous student strikes with a view to obtaining better living and study conditions, the government attempted to alleviate this situation by promulgating and implementing a decree organizing the operation of student associations (Decree No. 82/83 of 19/2/82). In 1990, with the advent of the multiparty system, Cameroonian university students concentrated in the single state university of Yaoundé had become a major threat to the ruling party. They were in fact easy prey for the new opposition political parties which promised them much better living and studying conditions and a bright professional future. To deal with this situation, the government in place was obliged to create new universities to decongest that of Yaoundé. This solution was obviously based on the principle of "divide and rule" (Mbembe 1985; Socpa 2003; Bella 2015: 318–323). In the process, five new universities were created, including the universities of Yaoundé II (Soa), Douala, Dschang, Ngaoundéré and Buea. In addition, the former University of Yaoundé was renamed the University of Yaoundé I (Decree N° 93/027 of January 1993).

Preceding the 1993 reforms a debate took place within the Ministry of Higher Education on the employability potential of the disciplines and curricula taught at the University of Yaoundé. The objective was to provide students with skills and experiences that could help find solutions to the unemployment crisis. Thus, each discipline was invited to reflect on the possibility of reforming the university system with a focus on capacity building and skills acquisition. In this context, the anthropology curriculum was redesigned with new specialisations to provide students with a basic understanding of the discipline's applicability and practice.

The status of anthropology in academia before the 1993 reform

Since the creation of the Federal University of Yaoundé, some anthropology courses have been taught at the Faculty of Humanities and Social Sciences (FALSH). Designed according to the French university system, anthropology (following the French tradition) was taught as a sub-discipline of the social sciences. All sociology students were required to take courses in anthropology. Although there was no separate department, the study of anthropology was an essential component of the degree in sociology.

The decree creating the Federal University in 1962 stipulates in its article 11 that the Faculty of Letters and Humanities shall award Bachelor of Arts degrees to students who have passed the examinations leading to the award of the five higher education certificates that make up the Bachelor's degree (Republic of Cameroon 1974; Republic of Cameroon 1982). The following degrees were to be awarded: Classical Literature, Modern Literature, History and Geography, and three Humanities degrees: Philosophy, Sociology, and Psychology.

To obtain a degree in sociology, students had to receive an anthropology certificate, as the sociology degree comprised the following certificates:

- Certificate in the History of Civilizations,
- Certificate in General Sociology,

- Certificate in General Psychology,
- Certificate in Anthropology,
- Certificate in Political and Social Economy.

From the beginning, conscious efforts were made to integrate anthropology in the first years of the Federal University of Cameroon. The following courses were included in the sociology curriculum under the heading "History and Methodology":

- Major theories in sociology (*ethnologie*),
- Definition of sociology and social sciences,
- Sociology and psychoanalysis,
- Sociological research, field research methods and techniques,
- *Ethnologie* and sociology, the significance of archaism.

The reforms of the 1980s merged the three humanities degrees into one philosophy degree. These reforms proposed to remove sociology and anthropology entirely from the curriculum, a move strongly resisted by sociologists and anthropologists, whose action led to the creation of a separate sociology department at the University of Yaoundé in which both disciplines were taught. Indeed, these reforms allowed students to take joint courses in philosophy and sociology during the first two years of their studies. In the third year, they specialised in sociology while obtaining a combined philosophy/sociology degree. Anthropology courses were reduced to a single course entitled Sociology of Traditional Africa. Students interested in anthropology could only start taking basic courses in the third year, with the possibility of concentrating on anthropology in the fourth and fifth years. The courses focused on the anthropology of Cameroon and the world, but also included courses with an applied orientation, such as urban anthropology, applied anthropology, and anthropology of development.

Although the students obtained degrees in philosophy-sociology, they were taught basic anthropology because most teaching staff were anthropologists who had been trained abroad notably in France, Switzerland, Belgium, Germany and the United States of America. Yet because anthropology had no visibility, most trained anthropologists preferred the academic profile of sociologists or historians (Nkwi and Socpa 2007; Fokwang 2008).

The status of anthropology teaching from 1993 to 2023 in Cameroonian universities

The mission of the discipline

Anthropology as an academic discipline focuses on human biological and cultural variation in time and space. It has six subfields across different universities in Cameroon: social or cultural anthropology; biological or physical anthropology; archaeology; linguistic anthropology; medical anthropology; and anthropology of development. The mission is to increase and disseminate knowledge of anthropology through empirical research, teaching and publications. The goals and obligations are to reach and educate our primary constituencies (graduate, masters and doctoral students) and the wider public. The ultimate goal is to routinely address, associate, cooperate and collaborate with students within and outside the Universities. The development sector has been identified as the primary target job market for anthropology students.

Employment opportunities

With the 1993 university reform, anthropology – in common with other disciplines – was required to prepare students for the job market. Graduates are trained to create self-employment jobs as international consultants and researchers. They are prepped to be employed in national and international organisations and NGOs in competing positions with students from other universities, professional schools, and other countries.

Graduation requirements at the Master's and Doctoral level

Each candidate is attributed a supervisor of the thesis by the Department. Candidates are required to present a research thesis at the end of the programme. The thesis needs to make a scientific contribution to research by handling original topics and providing a critical analysis which aims to solve critical socio-anthropological problems and to contribute to human and environmental development. This implies that candidates focus on their original proposals and topics of research by building on the academic coursework, workshops, seminars, internship facilities and research experience so as to produce critical analyses and contributions in their theses. The thesis is submitted three to six months after the coursework has been completed and successfully validated.

Each candidate is expected to present two to three seminar papers before completing the courses and the final thesis. This enables the candidates and their supervisors to appreciate and devise critical evaluations and suggestions for improvement of the research and critical analysis of the work and other future works of the candidate.

Candidates need to earn at least a C grade, that is, a Grade Point Average (GPA) of two (2.0) points out of four (4.0), to be eligible for graduation. Unsuccessful candidates are given two more chances to graduate. This means that the maximum period for graduation for a candidate is four years. The Master's degree shall be awarded to candidates who have successfully completed the course requirements and defended a dissertation from such research work before a panel (Kekedi 1979). Graduates must have completed a total of 120 credits (2/4) which allows them to register in the PhD/Doctorate cycle.

Anthropological curricula

The anthropology curricula in Cameroonian universities are structured around standard core courses and specialised degrees. The curricula are developed by a commission of senior ranked anthropology staff of the universities who are encouraged to consult the teaching programmes of other universities in the world, but also to propose courses that are aligned with local realities. The standard core courses are generally the same in all universities that offer anthropology and include the following:

- Introduction to general anthropology
- History of anthropology
- Research methods in anthropology
- Cultural dynamics
- Kinship systems
- Belief system
- Culture and personality
- Current theories in anthropology

The specialisation courses differ by university, depending on the specialisations of the senior staff. Yet, given the focus on applied anthropology, many universities offer anthropology of development and medical anthropology as possible specialisation degrees.

Table 1: Cameroonian universities and their specialisation courses

Specialisations	Universities
Cultural anthropology	University of Yaoundé I University of Douala University of Maroua University of Bertoua University of Ngaoundéré
Anthropology of development	University of Yaoundé I University of Douala University of Maroua University of Buea University of Ngaoundéré
Medical anthropology	University of Yaoundé I University of Douala University of Maroua University of Ngaoundéré
Visual Anthropology	University of Maroua University of Ngaoundéré
Social anthropology	University of Buea

Evolution of anthropology department staff in Cameroonian universities today

When the Sociology Department was created in the 1980s, Jean Pierre Warnier, a French anthropologist and specialist of Grassfields societies in Cameroon, served as its chairman for twelve years. By 1986, the sociology department had twelve staff members, including seven anthropologists. In 1993, the discipline was given autonomous status as a section in the new Department of Sociology and Anthropology and was entitled to award degrees (B.Sc., M.Sc., PhD) in anthropology.

In 1986, almost all staff members were trained abroad. Only two had been trained entirely in Cameroon. The others had received their degrees from universities in the USA, France, Switzerland, Germany and Belgium and came from various countries. While this was a great asset to the department, they returned to a university system that offered them little incentive for research and teaching. The lack of resources further aggravated the situation and negatively impacted their careers which seemed to stagnate once they became lecturers.

As a result of the university reform of 1993, the number of public universities in Cameroon has increased from one (Federal University of Yaoundé) to eight, with an additional four being created in 2022. As we complete this article (December 2023), there are autonomous anthropology departments in three public universities (University of Yaoundé I, University of Douala, University of Ngaoundéré) as well as three anthropology sub-departments in the universities of Buea, Maroua and Bertoua. The anthropology teaching staff at these six universities comprises 20 professors of magisterial rank, including five full professors and 15 associate professors. While a good number obtained their PhDs abroad (France, Belgium, Netherlands, Great Britain), several were trained at the Department of Anthropology of the University of Yaoundé I, which renders this Department a pillar in the training of anthropologists in Cameroon. By 2022, the following specialisations were represented at the level of associate and full professors in the anthropology departments in the different universities (descending in order of frequency): cultural anthropology, medical anthropology, development anthropology, social anthropology, food anthropology, and religious anthropology (see Appendix).

Moreover, the anthropology department in Yaoundé has also been instrumental in supporting anthropology departments in other universities. Indeed, to make up for the lack of teachers in the newly created universities, anthropology lecturers from Yaoundé are constantly invited to participate in inter-university mobility and to lecture and serve as jury members for master's and PhD examinations (Ekambi and Saïbou 1993).

Inter-university cooperation, also between universities in Cameroon and other parts of the world, is very intense. For example, the Anthropology Department of the University of Yaoundé I entertains several collaborations with universities in France, Germany, Japan, the Netherlands and Switzerland. While some are formalised by framework cooperation agreements and aimed at capacity building on the level of staff and students, others are more informal and rest on the initiative of individual staff members. The following table gives a brief overview of the international collaborations of the Anthropology Department of the University of Yaoundé I over the past twenty years and the staff members who have benefited from these (see table 2). These examples and many others attest to the importance of inter-university cooperation for Cameroonian students and lecturer-researchers.

International inter-university cooperation: the experience of the anthropology department of the University of Yaoundé I

Currently, anthropology training curricula in Cameroonian universities emphasise applied issues. Beyond theoretical training, practical training aims to provide professional skills that transform students into cultural brokers for development and consultants for organisations seeking their expertise. The training is tailored to the requirements of the end-users: government, national and international organisations. Applied areas include medical and health issues, development issues, systems analysis, urban development, local development and environmental issues.

Experience has shown that international organisations and NGOs are calling upon more and more anthropologists to help bring about meaningful change in rural communities. The Cameroonian government, intergovernmental agencies and international NGOs, such as the German Agency for International Collaboration (GIZ), the International Union for Conservation of Nature (IUCN), the Netherlands Development Organisation (SNV), the French Association of Volunteers for Progress (AFVP), CARE International, as well as several UN agencies (e.g. UNDP, UNESCO, UNICEF, WFP, WHO) and many other international organisations, have called on anthropologists. These development partners have often offered anthropologists better working conditions and more appropriate remuneration than the academic environment in Cameroon.

Table 2: International inter-university cooperations at the University of Yaoundé

Partner Institutions	Projects	Partners/Spin-offs
Wenner Gren Foundation for Anthropological Research (USA)	Funding of the annual conferences of the Pan African Anthropological Association (PAA)	Creation of networks: NAWA (Network of African Women Anthropologists) NASA (Network of African Students Anthropologists)
Leiden University (The Netherlands)	Funding of doctoral studies	PhD studies of five former students of University of Yaoundé I

Partner Institutions	Projects	Partners/Spin-offs
University of Zurich (Switzerland)	Funding of doctoral studies	Partner: Prof. Tobias Haller PhD studies of one former student of University of Yaoundé I
	Funding of research and teaching collaboration	Partner: Dr. Michaela Pelican Collaborative research and teaching programme benefiting MA students of the Department of Anthropology, University of Yaoundé I One PhD thesis
University of Frankfurt (Germany)	Funding of three PhD theses	Partner: Dr. Ute Röschenthaler Two PhD theses
University of Cologne (Germany)	MoU between the University of Yaoundé I and the University of Cologne Funding of research and teaching collaboration	Partner: Prof. Dr. Michaela Pelican Capacity building of lecturers: Training programme in grant writing. Collaborative research and teaching programme involving six students from the Universities of Yaoundé I, Dschang, Bamenda and six from the University of Cologne[1]
Musée de l'Homme (France)	Student supervision	One PhD thesis defended in the framework of formal cooperation between the UYI and the Museum of Man
University of Kyoto (Japan)	MoU between University of Yaoundé I and Kyoto University	Partners: Prof. Otha (KU) Prof. Misa Hirano-Nomoto (KU) Prof. Maurice Aurélien Sosso (UYI)
Tokyo University of Foreign Studies (Japan)	MoU and Agreement of cooperation and exchange between University of Yaoundé I and TUFS	Partners: Prof. Kayako Hayashi (TUFS) Prof. Takanori Oishi (TUFS) Prof. Maurice Aurélien Sosso (UYI)

1 See Pelican and Ngeh in this volume.

Which way to go?

The future of anthropology in the coming decades will depend on how the discipline addresses the challenges of the 21st century. One of these significant challenges is to meet the needs and demands of a rapidly changing world. The rehabilitation of the discipline's image is well on the way to full recovery. Still, the integration of anthropological knowledge into the development process of African nations must become a permanent feature. The 1990s opened a window of opportunity and only quality research and training will keep that window open. If action research is of paramount importance to policy makers, the work of applied anthropologists must feed into the teaching programmes within the academy.

If the teaching of anthropology in Cameroonian universities is experiencing a significant increase in the number of students, this is mainly due to the vision that anthropology as an applied discipline can offer jobs in the private and public sectors that seek to address developmental issues. To do this, curricula must address theoretical, conceptual and practical needs relevant to employability.

Indeed, anthropology can only gain respectability among the social sciences if its performance as an applied discipline meets the standards. Such an achievement will depend on the commitment to solid training in theory and practice. Anthropology's survival will also depend on its ability to both strengthen and transcend the micro-analytical approach and reach a more interdisciplinary and transdisciplinary understanding. The more students are trained in comparative, multidisciplinary methodology, the better equipped they will be to work with others whose knowledge and understanding of anthropological ideas is limited or biased. Anthropologists must learn to work in multidisciplinary teams because the development problems of today and tomorrow are no longer the concern of a single discipline. Only by opening up to inter- and transdisciplinary collaboration can anthropology assert itself as a strategic discipline in the development equation.

Conclusion

The experience of teaching and practising anthropology at the University of Yaoundé I has served as a model for the newly founded Cameroonian universities. Thanks to cross-disciplinary teaching, many students enrolled in other fields of study are interested in anthropology. The training of this new generation of students will undoubtedly help develop a new advocacy and marketing of anthropology beyond its disciplinary confines. As our students enter certain professions, such as the military, the police, and private entrepreneurship, one can hope this will create a niche for the discipline in the future.

The employment of our students by non-governmental organisations to carry out various tasks opens another window of opportunity for the discipline. As NGOs are increasingly involved in grassroots development activities, they will continuously need anthropological knowledge. If training is tailored to the practical needs of development organisations, the academy will strengthen its role and place in the development equation. It has been said that the lost decades of African development are due to the neglect of the micro-analytical approach, and it is hoped that anthropology will provide the necessary knowledge to make a difference.

In the age of globalisation, Africa faces the dilemma of confronting cultural invasion and the revival of African cultural values. Many nation-states will ultimately face the difficulties of straddling African and foreign cultures, borrowing what is essential and retaining what is beneficial. Africa's renaissance will depend on its ability to confront invasive cultures whose technological advantage will undermine African values and norms. If African nations are to preserve cultural values and norms now and in the future, they must invest in strengthening and employing anthropology students. The recent reforms in Cameroon that have enabled anthropology to become a discipline in its own right are commendable.

Acknowledgements

We thank Heike Becker, Michael Bollig and Michaela Pelican for their insightful comments and editorial suggestions to earlier drafts of this paper.

Bibliography

Asad, T. (ed.) (1973): Anthropology & the Colonial Encounter, Ithaca Press.

Beals, R./Hoijer, H. (1965): An Introduction to Anthropology, New York: McMillan Co.

Bella, A. E. (2015): "Les mouvements étudiants et l'ordre politique au Cameroun postindépendance: entre enrégimentement et resilience." In: Syllabus review 6/2, pp. 291–328.
De Haas, M. (1994): Anthropology, Dept of Anthropology, University of Natal.
Ekambi, M.J./Saïbou, N. (1993): "Répertoires des mémoires et thèses en Sociologie et anthropologie. " In: Annales de la Faculté des Lettres et Sciences Humaines, Université de Yaoundé. Faculty of Letters and Human Sciences, 1978, Livret de l'Etudiant, 1978–1979, Yaoundé.
Fokwang, J. (2008): "Themes and Legacies: Anthropology's Trajectories in Cameroon." In: A. Boskovic (ed.), Other People's Anthropologies, Oxford: Berghahn Books, pp. 125–141. https://doi.org/10.1515/9780857450203-010.
Goodenough, W. P. (1970): Description and Comparison in Cultural Anthropology, Chicago: Adline Publishing Co.
Keesing, R. (1981): Cultural Anthropology: A contemporary Perspective, New York: Holt, Rinehart, and Winston.
Kekedi, N.R. (1979): Les étudiants de l'Université de Yaoundé, Dissertation for the University of Yaoundé examination. Maîtrise, University of Yaoundé, pp. 137.
Marchand, C. (1962): La scolarisation française au Cameroun, ORSTOM.
Mbembe, J. A. (1985): Les Jeunes et l'ordre politique en Afrique, Paris: L'Harmattan,
Nkwi, P. (2006): "Anthropology in a Postcolonial Africa: Survival debate." In: G. L. Ribeiro/A. Escobar (eds.), Global Anthropologies, Disciplinary Transformation within Systems of Power, Oxford, New York: Berg Publications, pp. 157–180.
Nkwi, P.N./Socpa, A. (2007): "Anthropology at the University of Yaounde I: A Historical Overview, 1962–2008." In: The African Anthropologist 14/1&2, pp. 65–88. https://doi.org/10.4314/aa.v14i1-2.57730.
Republic of Cameroon (1974): Council for Higher and Scientific Education and Technical Research, Yaoundé.
Republic of Cameroon (1982): Law on Higher Education, Science and Technology.
Republic of Cameroon (1993): Réforme universitaire du Cameroun, Yaoundé, 351p.
Socpa, A. (2003): Démocratisation et autochotonie au Cameroun. Trajectoires régionales divergentes, Münster: Lit Verlag.
The University of Yaoundé (1978): Student's Guide, 1978–1979, 61p.

Decrees

Decree No. 82/83 of 19/2/82 fixing the roles of the organisation, the functioning and the management of the company, and the control of the students' associations in the faculty of the Grandes Ecoles.

Decree No. 93/027 of January 1993 creating new state universities in the Republic of Cameroon

Appendix

Anthropologists of magisterial rank in Cameroonian universities between 1993 and 2023

N°	Names	Institutional Affiliation	Specialisations
		Rank of Full Professor	
1	Nkwi Paul Nchoji	University of Yaoundé I	Medical anthropology Cultural anthropology
2	Mbonji Edjenguèlè	University of Yaoundé I	Cultural anthropology Development anthropology
3	Socpa Antoine	University of Yaoundé I	Medical anthropology Social anthropology
4	Mebenga Tamba Luc	University of Yaoundé I	Cultural anthropology
5	Kum Awah Paschal	University of Yaoundé I	Medical anthropology
		Rank of Associate Professor	
6	Njikam Margaret S.	University of Douala	Medical anthropology
7	Ngima Mawoung G.	University of Yaoundé I	Food anthropology
8	Essi Marie José M.	University of Yaoundé I	Medical anthropology
9	Ndjio Basile	University of Douala	Cultural anthropology

N°	Names	Institutional Affiliation	Specialisations
10	Moussima Nganjo	University of Ngaoundéré	Anthropology of art
11	Akoko Robert MBE	University of Bamenda	Anthropology of religion
12	Edongo Ntede P. F	University of Bertoua	Cultural anthropology
13	Ndzana Bertrand	University of Ngaoundéré	Development anthropology
14	Deli Tize Teri	University of Yaoundé I	Development anthropology
15	Afu Isaiah Kunock	University of Yaoundé I	Development anthropology
16	Ismaela Datidjo	University of Dschang	Cultural anthropology
17	Otye Elom Ulrich	University of Ebolowa	Cultural anthropology
18	Wogaing Jeannette	University of Douala	Social anthropology
19	Abouna Paul	University of Yaoundé I	Cultural anthropology
20	Ngo Liken Julienne	Ecole des Sciences et de la Santé (ESS – UCAC)	Medical anthropology

Notes on the Institutional Development of Anthropology in Egypt, and Toward its Decolonisation

Daniele Cantini

Introduction[1]

Egypt has provided inspiration for Western anthropologists almost from the discipline's inception; and with some significant variations over different historical periods, it has continued to do so ever since. Travellers who wrote accounts of their journeys describing the customs and manners observed along the way (for example, Edward Lane or Johann Ludwig Burckhardt) exerted an enormous influence that lasted for decades. Such accounts affected the first Egyptian scholars who tended to study Bedouin populations and focus on "primitive" and remote communities, and on cultural artefacts that were in danger of disappearing because of the encroachment of the modern way of life. At the turn of the 20th century, ethnography was already well established as a discipline and as a distinctively modern mode of analysis in the country, with two separate branches of the discipline enjoying some prominence: the research in biological anthropology, which among other things tried to situate the Egyptian "race" between Europe and Africa (Boëtsch 1995); and the researches being carried out in partial cooperation with the Société Géographique d'Egypte, founded in 1875, with the aim of collecting the folklore heritage of the different peoples and tribes across the country, including the Sudan (Perrin 2005; El-Shakry 2007). Almost all anthropologists

1 Some parts of this chapter also appear in Daniele Cantini and Amal Abdrabo, "Tracing Histories and Institutional Developments of Anthropology in Egypt" in Daniele Cantini, Abdallah Alajmi, Irene Maffi and Imed Melliti (eds.) (2025). *Social Anthropology in the Arab World: the Fragmented History of a Contested Discipline*. New York/Oxford: Berghahn.

associated with these trends were Europeans. While this period was marked by a European predominance, ethnology did not come to Egypt as a direct result of the colonial experience, but rather as a complex endeavour in which different logics were at play, including the desire to modernise the country.

Social anthropology came at a later stage, during the British colonial occupation of the country. Egyptians started to be attracted to anthropology in the 1930s, with a group of scholars who trained abroad and then carried out fieldwork in Egypt or the Sudan (Hopkins 2014). By this time, the colonial-modern mode of social-scientific inquiry had established itself in Egypt, as elsewhere, bringing forth new understandings of "society" and "progress", for instance, and indeed a new understanding of knowledge and of its relation to society. This mode of knowledge production had transformed Egypt into a "great social laboratory", in which the peasantry, or indeed the entire population, became the objects of scientific inquiry, understood as social engineering, with very little change occurring in the post-independence phase following the colonial period (El-Shakry 2007). Strategies of governance based on the development of instrumentalist knowledge, statistical languages, the logic of rational planning and, more broadly, the systematic targeting of the subaltern population toward improvement in the fight against backwardness were all features early on. During these, the insistence of colonial anthropology, ethnographic or anthropometric, on race became an insistence on identity, less interested in colonial difference and more interested in the uniqueness of the collective national subject (ibid.). There was not a simple reproduction of colonial practices and understandings, but rather "an attempt to render models of modernity intelligible through the grid of indigenous social and cultural values and practices, and through reformulation or critique" (El-Shakry 2007). The question of how to be modern while maintaining the specificity of cultural identity was a central preoccupation of Arab intellectuals in the 19th and 20th centuries. El-Shakry notes how in resisting the totalising and racialised nature of European claims to progress, reason, and the nation-state, the Egyptian nationalist intelligentsia staked the claims of social science on the particularity of local difference – as in the attempt to create an "Arab social science". In attempting to argue, however, that as non-Westerners Egyptians had internal indigenous sources of progress, indigenous reformers inadvertently accepted many of the very premises central to western categories of thought (progress, reason, the nation-state).

An indigenous anthropology was hard to come by, however, for a variety of reasons. Particularly in the 1940s and 1950s, at the height of the nationalist struggle, anthropology had a very marginal role, probably since it was seen with scepticism as an instrument of colonial domination, suspected of primitivising Egyptians (Hopkins 2010: 5). Other disciplines were deemed to be more attuned to the time, and more relevant for the development of the Egyptian nation, for instance, economics, sociology, and political science, obviously in addition to the "exact" sciences and medicine. This trend continued after independence and was surely exacerbated by the fact that local institutions were increasingly understood as instrumental in reaching development goals set by the different governments. Social sciences as a whole came to be considered useful in solving societal problems, an understanding that proved durable. As I discuss in this chapter, this understanding of social sciences as being at the service of the state, and instrumental in allowing developmental projects to be implemented, is clearly a central feature in the shaping of the conditions of possibility of anthropology in Egypt, as elsewhere in the region.

In this early nationalistic phase, the few anthropologists who were active in Egypt had been trained abroad, mostly in the UK and less so in France and the USA. The most prominent among them is perhaps Sayyid 'Uways, and Ahmed Amin if we include folklorists. Their research, largely focused on finding out the authentic character of Egyptian people, was partly in line with the constructivist approach to knowledge and to the need for nation building (Roussillon 1985). In this phase, however, there was still no institutional base for Egyptian anthropology, if we exclude the American University in Cairo (AUC) and its Social Research Centre (SRC). Another two decades elapsed before the founding of the first department of anthropology, at Alexandria University in 1974; in this phase, there are a few towering figures who established the discipline in Egypt, and who perhaps inevitably, largely determined its shape. Among them, Ahmed Abou Zeid is widely considered to have been the most influential. At the time of writing, there are six departments in public universities all over Egypt offering degrees in anthropology.

Knowledge produced locally has had a very limited influence on international academic discourse even when it directly concerns the region. This characteristic was noted quite early on "a large part of the reason lies in the historical development of anthropology as a discipline and the nature of relations between the First and Third Worlds. These issues have been insightfully treated, for the Middle East, by Asad, Said, and Turner" (Shami 1989). Without negating this underlying reality, Seteney Shami focuses on the socio-cultural anthropology produced within the Arab world. Such analysis requires an understanding of the local development of university education and the social sciences generally. In this chapter, I offer some preliminary observations to continue such work.

Anthropology in Egypt is marginalised as a discipline within the country, being simultaneously mocked as "folklore" within the academy and feared for the proximity to research subjects that the ethnographic method requires. Moreover, this happens within the context of a widespread crisis of education, higher education and research systems that is well known and debated both inside and outside the country. However, despite all these difficulties, particularly after 2011, there was a resurgence of interest in anthropology as a discipline among young Egyptians, who – in contrast to their great-grandparents, who were put off by the idea that anthropology was a colonial science dealing with "primitive" people – were attracted by the closeness to the street and the people that anthropology, and particularly the ethnographic method, promised.

In this chapter, I offer an overview of the development of anthropology as a discipline in Egypt. I first discuss the different phases of this development, before focusing on the institutional dimension. I then problematise the cleavage between anthropology *on* Egypt and anthropology *based in* Egypt, by discussing some authors who, particularly in the 1990s, tackled this question, often when moving in-between worlds. I thus go back to the institutional dimension in recent years, paying particular attention to how collaborative projects attempted to readjust the imbalance between local and international production of knowledge. I conclude by offering some glimpses of what has been happening since the 2011 revolution.

An Egyptian anthropology vs. the anthropology of Egypt?

This brief overview of anthropology in Egypt should by no means be taken as an assessment or an account of the state of the art in the development of a regional variety of the discipline, a task for which I would be greatly inadequate. As I will make clear in what follows, it is not immediately apparent how to differentiate between an indigenous and exogenous anthropologist, given the number of Egyptians who were educated abroad, and the different layers of interdependence, for example, at the level of research funding. Similarly, Nicholas Hopkins (2010) in detailing the origins of anthropology in Egypt does not attempt to define a differentiation because his analysis ends before the institutionalisation of anthropology in Egypt. Moreover, as pointed out by Jean-Noel Ferrié in his introduction to a two-volume examination of anthropology of Egypt, the very idea of an Egyptian anthropology, or the interest in detailing its state of the art, should not be taken as self-evident, "*ce ne va pas de soi*" (Ferrié 1995). In his critical examination, he claims that it would be necessary to be explicit about what the goal of such an assessment would be, to avoid that the analysis of the works produced by Egyptian anthropologists transforms them "more or less consciously from producers to objects of knowledge or, in the best cases, unwilling informants" (ibid., my translation).[2] Presenting the works collected in the two volumes, he goes on to specify that most of them are studies on Egypt, where Egypt is the field, although two authors included in the collection are indeed Egyptian. "We do not pretend to assess the Egyptian anthropology, that is an anthropology produced in Egypt by Egyptians trained in Egyptian institutions, the essence of which production would pass through special needs, also defined as Egyptian" (ibid.). Such an endeavour, he concludes, pertains rather to the sociology of science, and in any case the interest of such an exercise is not to be taken as self-evident. However, this should not be construed as a lack of attention. Ferrié indeed recognises that it would be difficult to examine Egyptian anthropology, because it is not well developed, and, he adds, the causes of such lack of development are to be researched within the structure of the academic field in the country, rather than in the oft-repeated points about the link with the colonial past or because it does not sit well with developmental goals.[3] Apart from Ahmed Abou Zeid, who "took part in the reformation of a Mediterranean anthropology at the beginning of the sixties", "not many names come to mind" (ibid.) if one considers the impact they had on the international community. One could surely not speak about an academic community nor an established anthropological tradition.

The debate on decolonising knowledge offers the possibility for a different approach, and this became apparent in Egypt particularly after the 2011 revolution. Issues of who produces knowledge, on what, and from where, have been a necessary step in producing anthropology all over the world since at least the crisis of the discipline in the 1960s. Inequalities in access to research training and facilities, as well as in being able to access the international publishing markets, have always existed, but these are increasingly being questioned,

2 As elsewhere in this chapter, all translations are mine.

3 The flourishing of Indian anthropology, in his analysis, proves the point. The lack of development in Egyptian anthropology must be investigated within Egypt and its academic field.

and rightly so. Moreover, the ideal of academic cooperation is becoming central in the new understanding that anthropology has of itself; in this context, a closer examination of the local conditions of producing knowledge and research is definitely not an assessment in the sense of making an inventory of what is lacking, but rather the precondition for more adequate future research projects. In my view, and as I have already started doing, research projects should include as much training as possible, not only in the Global North but also in Egypt, with the aim of familiarising foreign researchers with the specificities of doing fieldwork in, and producing knowledge on Egypt. A discussion of the local developments of the discipline within the broader political and social contexts, as well as in relation to the developments of anthropology globally, thus does not have the goal of evaluating or assessing, but instead aspires to instigating open discussion, with the explicit aim of trying to overcome contemporary inequalities, as I have made clear elsewhere (Cantini 2021).

Finally, I do not want to imply that knowledge produced by Egyptians on Egypt is more authentic, or less in need of explicit position-taking than scholarship produced by non-Egyptians working on Egypt. All scholarship needs to be situated: it cannot be taken as given and needs to be critically assessed. Decolonising knowledge does not necessarily mean dismissing all that is produced elsewhere (for an earlier take see Morsi et al. 1990, which I discuss below), but should instead aim to increase our knowledge and understanding, particularly of the conditions that shape knowledge production. This has been my goal in much of the work I have done so far, and the importance of this approach seems to be confirmed by the rising interest around the discipline in Egypt, particularly after 2011.

The beginnings of anthropology in Egypt

Anthropology began to be practiced by Egyptians in the 1930s, and one of its earlier and more significant practitioners was Mohammed Galal (1906–1943). After completing high school in Zagazig, he went on to Paris to study; there he had the good fortune of studying under two of the most revered scholars at the time, Marcel Mauss and Luis Maussignon, finding himself working between different institutions that were just starting to grant degrees in anthropology. Thus "Mohamed Galal was arguably the first professional Egyptian anthropologist in the sense that he had academic training, carried out fieldwork, and published an anthropological study. Unfortunately, he died prematurely. His early death and the relative inaccessibility of his major publication mean that he is largely unknown to the present generation of social scientists in Egypt or the Middle East" (Hopkins 2014: 175). His major publication was a book-length study of funerary rituals in contemporary Egypt, which his mentors feared could rouse the ire of the country's religious scholars; hence the suggestion to publish his thesis in France, and to consult Taha Hussein – an influential figure in Egyptian intellectual history – for some key rewordings (ibid.).[4] For this book, he conducted research in different areas of Egypt, particularly Sharqiya, but also in Beheira, Asyut, Aswan, and Cairo. During his research, he sought direct contact with the rural population, avoiding the "sanitized view" that officials would have provided for him (ibid.: 179).

During his years in Paris, he also helped to curate two ethnographic collections for the Musée de l'Homme. His mentors helped him administratively and financially, finding a way for him to receive a stipend from Egyptian authorities; this was however paid irregularly, and Mohamed Galal was denied support for his subsequent fieldwork in Sudan, or for publication of his thesis (ibid.). In 1938, Galal was finally awarded a grant from the Institut d'Ethnologie for research in Sudan. The grant was a significant amount of money, and apparently comparable to that which French researchers under the auspices of the Institut d'Ethnologie were receiving. He undertook more than a year of fieldwork in different parts of South Sudan, having also sought advice in Oxford from Evans-Pritchard and Seligman. Once war broke out, Galal was evacuated by the colonial authorities to Egypt, where he eventually taught at Cairo University. He then contracted an inoperable brain tumor, died, and was buried in the Sendenhour cemetery in 1943, leaving a wife and two daughters (ibid.: 186–187).

Hopkins notes that Galal's pioneering efforts were not followed up in Egypt because of his early death, and because of the interruption occasioned by the war. The fact that his main written work was published outside anthropology

4 Only two decades earlier, Mansour Fahmi, an Egyptian scholar who had studied in France and published his philosophy thesis on the condition of females in Islam underestimated the wrath of religious scholars at home. As a consequence, his professorial position at the Egyptian University was cancelled, and he had to wait years before being reinstated, also thanks to Taha Hussein. Subsequently, he never again published anything as controversial as his thesis. For more on Mansour Fahmi, see Reid 1990: 65–67.

in a journal of Islamic studies, and in French, may have discouraged successors from taking it into account. However, the decade of the 1930s in Egypt saw other social science beginnings. Sayyid 'Uways, who I discuss below, also began his studies at this time but did not publish until after 1945. While Galal's achievements cannot be taken as representative of the anthropology in Egypt of the time, his trajectory somehow exemplifies the pioneering phase of the discipline – largely individual efforts, with training abroad that often determined subsequent academic links, and a production that had almost no impact on Egypt due to the absence of infrastructures at home, such as academic jobs or scientific associations or journals.[5]

During this phase, another path, not limited to anthropology, was that of the amateur, the person with expertise in another field who would also embark on studies in ethnology and folklore. The figure who could be taken as representative is Ahmad Amin (1886–1954), who at the end of his career produced a singular work, a dictionary of the mores and customs of the Egyptians. His career was very similar to that of Taha Hussein. Both men came from a modest background, studied in the religious schools up to the tertiary education at *Al-Azhar*, followed by an encounter with modern culture. In Amin's case, however, this was done in an autodidactic way, without training abroad. After a career as a law professor at the newly established Cairo University, where he taught from 1926 until his retirement in 1948, he started working on his dictionary of Egyptian customs, traditions and expressions, which he published a year before his death, *Al-Qamus al-'adat wa al-taqalid wa al-ta'abir al-misriyya* (Perrin 2002). He presents this work as a collection of what he learned growing up in a traditional neighbourhood. This work does not provide a theoretical reflection on what traditions and customs are, and seems to be largely understood as having been inspired by the work of Edward Lane; moreover, in the framework of his larger work this singular reflection on folklore looks more like a *divertissement* than a serious disciplinary engagement (ibid.).

5 This also reflects in the lack of scholarship available in Arabic, on this phase and in general on the development of anthropology in Egypt. I thank Amal Abdrabo for this critical remark.

Writing at the service of society - anthropology and revolution

The period immediately after Egypt's independence saw, particularly from the late 1950s onwards, an expansion of education and higher education systems, probably one of the most visible results of the new system. While anthropology and the social sciences were really never at the top of the political agenda for scientific development, in this phase there was a significant institution building effort (Cantini 2020). For Egyptian anthropology, this was the phase in which more and more scholars went abroad, mostly to the UK but also to the USA, to earn their PhDs. Due to space limitation, I will only discuss one example here, Sayyid 'Uways (1913–1989) who was the most representative example of the generation who reached maturity in the 1940s and which was closely analysed by the late Alain Roussillon. This generation was a founding one, coming to assume and embody the "series of ruptures" that characterise contemporary Egypt – "confronted with the failure of the liberal experience and the exhaustion of the reformist thought [...] its members were responsible for assessing the colonial period at all its levels, from the economy and social structures to the system of values, in the bodies and minds of the people" (Roussillon 1985, my translation). Already evident to his contemporaries, one of the greatest merits of Sayyid 'Uways was to be able to inscribe these ruptures within the continuity of what he called "the eternal Egypt", while providing a model for integrating the changes brought forth by modernity into this structure.

A graduate of the newly established Cairo School of Social Work, 'Uways started working as a social worker mostly tackling juvenile delinquency. During this time, he was twice selected to attend non-degree programmes in the UK, until eventually entering a PhD programme at Boston University in 1952. In the thesis he wrote there he made comparisons between the youth delinquency occurring in problematic neighbourhoods in both Cairo and Boston (Hopkins 2010: 108). Upon his return to Cairo, he joined the newly founded National Center for Criminological Research, organised as a research unit under the Ministry of Social Affairs, where 'Uways had worked before embarking on his PhD.[6] He worked there as research professor until his retirement. Per-

6 Egypt's post-1952 leaders preferred to establish new institutions, such as the National Research Center (which had frail roots in the old regime) and the Center for Criminological and Sociological Research, rather than to refurbish old ones like the Egyptian Geographical Society (Reid 1993) or to invest in universities as research institutions,

haps his most famous work is a book in which he analyses letters written in the 1950s to the Iman al-Shafi'i, a prominent Sunni scholar who lived in the 9th century and was buried in Cairo (ibid.). In his work, he constantly mediates between the present society and its past, the tradition, the authentic, so as to make apparent the deep differences within the present and at the same time show the strength of the links that bound it to the past (Roussillon 1985). But he also mediates between the inner truth of society and the external world, and between national cultures and western values in the process of globalisation.[7] His output is characterised by a push towards scientific, socio-anthropological research from one side, and from the other by a sort of Romanesque writing. According to Roussillon, Sayyid 'Uways exemplifies the "figure of the organic intellectual, capable of self-identification with collective aspirations and of formulating a model of mobilization that allows at the same time for social change and for fidelity to itself" (ibid.).

After the 1952 revolution Egyptian universities underwent a serious transformation, partly dependent on the massification of access, a key promise of the revolution itself particularly when linked to employment opportunities in an expanding public sector, and partly dependent on the new ideology of the state. The university was regarded by the army officers as a stronghold of conservatism and anti-revolutionary thought (Najjar 1976). In 1954 over 60 professors lost their positions in what became known as the Nasserite purges (Reid 1990), and the revolutionary powers exerted considerable effort in transforming academia toward their goal of building a new society (Awad 1963). Particularly after 1961, Nasser intensified the pressure on universities to conform to the socialist design of the revolutionary society. Universities however were never completely under governmental control, not even at the height of Nasser's power; the regime sought to establish research centres

---

since they were less easily brought under state control (see Cantini 2021 for a detailed discussion). Some of these research institutes have very interesting histories, rather telling of political developments and of fashionable topics in research. A proper history of anthropology in Egypt cannot be written without examining their role, alongside those of universities.

7 A similar preoccupation with not only accounting for Egypt's uniqueness but also for its belonging to the region, and to Islam, originated the multivolume "*Shakhsiyyat Misr*" (*The Character or the Personality of Egypt*), in which Jamal Hamdan, a geographer who became a full-time writer, swept through millennia of Egyptian civilisation from an historical geographical perspective (Hamdan 1970).

instead, to promote younger academics deemed more loyal to the new ideological line, and to pursue interests more attuned with political goals (see note 6). Some institutes survived from the previous phase, adapting to the new political circumstances; research in this endeavour was clearly subordinated to political needs. For example, the Institute of Sudan Studies, hosted at Cairo University and created by King Faruq – who overprinted "King of Egypt and the Sudan" on postage stamps – was made independent from the University and renamed The Institute of African Studies in 1955 when the Sudan moved toward independence from Egypt as well as from Britain (Reid 1993: 564). The institute developed graduate programmes in geography, history, natural science, anthropology, politics and economics, and languages and dialects. In the 1970s, it began issuing a journal, the *African Studies Review*. By the 1980s, the *African Studies Review* no longer appeared regularly, and the institute now remains marginal to Egyptian national concerns. As long as Egypt's access to upstream Nile waters was not seriously threatened, the institute's place in Africa was not, after all, the prime concern of Nasser, Sadat, or Mubarak (ibid.). This interest seems now to be resurgent, and the institute was turned in February 2019 into a Faculty of African Graduate Studies, as a part of the plan to "serve the African continent", as stated in its mission.

The institutionalisation of anthropology in Egyptian universities

From an institutional perspective, the first department of anthropology was founded at Alexandria University in 1974; it was the first in Egypt and in the Arab world. Anthropology courses had been offered at Alexandria University since its inception in 1942, and the institution counted Radcliffe-Brown as one of the first professors, but this offering was often discontinuous, despite the presence of scholars such as Aly Issa, who guided several students between 1954 and 1964 (Hopkins 2010: 48–50). Anthropology was first taught at the American University in Cairo in 1956, and the Social Research Center, established three years before, hosted a series of mostly developmental projects in which several Egyptian researchers, including anthropologists, took part (more on this below). The situation was bound to change with the establishment of the first department entirely devoted to anthropology, and in a public university. In this context the pioneering figure was the one of Ahmad Abou Zeid (1921–2013).[8] He was the first to establish a school that spread across different Arab countries – he even claimed to have been the first to write "anthropology" in Arabic, in two articles about myths and "primitive" thought in 1946 (Hopkins 2010: 79). In the academic year 1999/2000, there were four full professors at the Institute for Anthropology at Alexandria University; one was Abou Zeid himself, at the time emeritus for more than a decade, and the other three had all studied under him (Lange 2005: 49).

Despite the department in Alexandria, and the few others established in recent years, anthropology in Egyptian universities is still however primarily affiliated with departments of the social sciences, especially sociology. Some of the social science professors working there contributed significantly to the shaping of the anthropological field in Egypt – for instance, Mohammed al-Gauhary from the Department of Sociology at Cairo University, and Alya Shukry from the Women's College (*Kulliyat al-Banat*) at Ain Shams University. A married couple, they obtained their PhDs in West Germany and went on to train vast numbers of anthropologists in Egypt (Lange 2005); the former was virtually the sole PhD supervisor for well over a decade (Dessouqi 2021). Courses in anthropology are now offered both at under- and postgraduate levels, and while the theoretical framework is largely structural-functional, there is an emphasis on the practical dimension of anthropology in order to implement modernisation and development plans and to help build the nation (Al-Sayed Al-Aswad 2006).

Ahmad Abou Zeid was supervised by Radcliffe Brown for his undergraduate studies at the University of Alexandria. In his memoirs he remarks that he was greatly influenced by his coming of age in the then cosmopolitan city of Alexandria, and of having been fascinated by French sociology before picking up an interest in anthropology. He then went on to Oxford, to achieve a B.Litt. in 1953 and then a D.Phil. in 1956, both under the supervision of Evans-Pritchard (Hopkins 2010). He thus studied under two of the main representatives of British functionalism, and this had a clear impact on how he thought of society – he regarded social structure as a unit, composed of several systems: ecological, economic, political, and kinship (Zayed 1995). He published widely

8 Transliteration of Arabic names could be inconsistent across different publications; this is also the case with Abou Zeid's name. Hopkins reports that Abou Zeid's preferred spelling was Ahmed Abou-Zeid, but in the Library of Congress catalogue he is spelled Ahmad Abu Zayd (Hopkins 2010: 79n). Other variations are possible as well.

in Arabic and English and his perhaps better-known book is on *al-tha´r* (feud or vendetta), the result of two months of fieldwork in a village in the Asyut district in Upper Egypt (Hopkins 2010).

In this context, it is perhaps more relevant to point to his legacy in establishing an understanding of what society is, and in setting the standards of anthropological research; both themes draw on fieldwork on which he published several handbooks (Lange 2005: 51). He was also very active on an institutional level, associating himself with the National Center for Criminological Research in Cairo, before becoming professor at Alexandria University, which he eventually led as Rector. He was a full member of the generation of the 1940s, briefly sketched above; a deep commitment to Egypt's development was to last through his life, as well as an attention to how change in values and orientation could be brought forth (Hopkins 2010).

This activism, however, did not prevent him from acknowledging the difficult condition of anthropology in Egypt and indeed in the Arab world, which in recent years he characterised as representing backwardness, *takhalluf* (ibid.: 52), largely for its failure in offering any theoretical advancement, limiting itself to becoming a data-collecting and largely ancillary branch of a discipline whose standards were continuing to be set elsewhere. In particular, he lamented the lack of a scientific institution or journal that could collect researches made in Egypt and put them into conversation with works from elsewhere. In 1995 there was a conference of Egyptian anthropologists, but its results are not easily accessible and it ultimately failed to reduce the dispersion of the practitioners (Lange 2005). I now turn to the discourse concerning crisis in education and research, before hinting at other non-university related developments in anthropology and then moving on to describe the present stance of anthropology in Egypt.

What is anthropology for? Anthropology as an academic discipline...

Abou Zeid's honest recognition of the dire condition of anthropology as a discipline matches a widespread sense of crisis, which the educational system as a whole – and as a consequence higher education and research – would have entered at some point during the 1970s, as I discuss in more detail elsewhere (Cantini 2021), and anthropology was no exception. This sense of crisis can be linked to many aspects, but here I would like to point mainly at the two which have been solidly established in the representation of the crisis over the last couple of decades: these are the crisis within the educational system on one side, and, on the other, the broader sense of political crisis, which included the subordination of research to politics and more recently to economic goals.

I dealt with the first aspect most directly in a research project I led on doctoral studies at Egyptian public universities, published in Arabic and in English (Cantini 2018, 2021); here I limit myself to a few considerations. Three decades ago, it had already been noted that public Arab universities faced serious shortages in funds, particularly as they had been confronted with massive growth since the 1970s. Professors, whose salaries are often meagre, are allowed, and at times actively encouraged, to increase their revenues by producing a textbook for each class they teach, which is then sold to the students. The resulting prioritisation of teaching over research makes the latter a limited exercise, particularly when field research is involved (Shami 1989). Since their inception the social sciences have been understood as a second-best option, and usually allow into their courses large numbers of students who were not "lucky" enough to gain admission to the more competitive departments of sciences and professional schools, to which they provide a cheap alternative, since they do not require labs or special equipment (Shami 1989). The establishment of social science departments is also related to another role that the university often fulfills in Arab society: that of a containing and even conservative force, tasked with providing youth with an alternative to unemployment (Hijazi 1986, quoted in Shami 1989). Despite some massive changes that the university systems in different Arab countries and in Egypt have undergone since the 1990s, when privatisation and internationalisation contributed to the enormous growth of the higher education system (Cantini 2016), this analysis is still by and large quite sound.

Secondly, in common with countries such as Iran or the Sudan, where anthropology has an indigenous history and development that make it necessary to discuss the issue of knowledge and power beyond a condemnation of coloniality (Shahshahani 1986, Monzoul 2018 as samples), in post-independence Egypt it was firmly established that science (including social sciences and hence anthropology) should be subordinated to, and at the service of, the state, a position which invariably conflated these disciplines with the executive power. The only way to justify the existence of such disciplines, generally regarded as luxuries, is by defending their value via the contribution they make to the solution of social problems (Shami 1989). This type of research found concrete expression in national institutions such as the previously mentioned National Center for Sociological and Criminological Research in

Cairo, which produced research output of a strictly utilitarian sort, with issues generally handled in an uncritical, ahistorical, and purely quantitative manner (Hijazi 1986, quoted in Shami 1989). The often-made parallel was that social scientists should understand their role as social engineers, and that sociology and anthropology are a sort of human technology; the goal of enhancing social control is never really too far away.

The specific problems of anthropology in this double impasse, within and without the university as an institution, are only exacerbated by its identification with the colonial time, and with its focus on "primitive" cultures. If science as a whole needs to contribute to the establishment of the nation, from a cultural and ideological point of view this nation is seen as arising from Islamic history and heritage and having a clear path ahead. Thus, a study of different practices and beliefs at local levels – a rather average pursuit for anthropology – is invariably seen as detracting from the desired overall unity: for example, a monograph focusing on family structure in an urban community was rejected by a major social science publisher on the ground that it did not contribute to an understanding of the unity of Arab identity (Shami 1989). This existential difficulty is made worse by some serious misunderstandings about the nature of the discipline, in part also caused by the superficiality of most research done by anthropologists who study groups that are safely marginal, such as Nubians and Gypsies in Egypt, Dinka in Sudan, and so on, invariably concluding that modernisation and increasing awareness of true Islam are inclining these groups towards better integration into society (ibid.: 653). As mentioned above, this harsh critique seems warranted by the overall emphasis on finding an authentic national identity after decolonisation, which not only in Egypt but, for example, also in Morocco, has been a core feature of the broader social sciences (Roussillon 2002). While these disciplines were originally imported during colonial times, and their methodologies and theories were largely adopted but not developed further, the disciplines were taken over by local elites and used to contribute to national constructs and to dispute the image of themselves that was provided for these societies by colonial science (ibid.). The hypothesis I advance here is that when the modernising efforts came to a halt, mainly for geopolitical reasons, social sciences also entered into a crisis mode, since their scope was no longer clear.

... and as a set of techniques serving social development

The crisis briefly sketched above should not be taken to imply that anthropology almost extinguished itself, as the opposite case was true. While, as an academic discipline, anthropology was too dispersed to be influential and not quite able to transcend the conditions in which it was founded in the post-independence period, from the late 1950s on there was a veritable boom of anthropological works as part of renewed developmental efforts, first under the auspices of the state, then funded by international donors. An anthropology at the service of societal change was high on demand in developmental projects, and there was no shortage of these in the newly independent Egypt, nor elsewhere in the world of course. Particularly during the 1960s and 1970s anthropologists, affiliated to the American University of Cairo (AUC) and national (Egyptian) universities and research centres, worked collaboratively in various development projects including the project of resettlement of Nubian communities initiated by Laila el-Hamamsy and Robert Fernea (Al-Sayyed al-Aswad 2006). The Nubian project is of particular relevance for several reasons; first, it took place at the height of Nasser's power, and yet it explicitly sought expertise from Western countries. It allowed collaboration between Egyptian and foreign scholars; this collaboration, mostly through the Social Research Center of the AUC, had already begun in 1954 with the analysis of voluntary resettlement projects of populations in the Fayoum and the Delta, and was extended in 1961 to the Nubian case (Hopkins 2010: 97). In the 1960s, the construction of the Aswan High Dam occasioned the forced displacement of a large part of the Nubian population. The project consisted in a survey of the Nubians to be moved and those already outside their historic homeland, with the goal of recording and analysing their culture and social organisation (Hopkins and Mehanna 2011).

Although not of this epic calibre, in subsequent years there was no shortage of developmental projects, despite all geopolitical developments. Although research in anthropology and the social sciences in Egypt depends on various funding sources including the national government, funding by foreign agencies and organizations, such as USAID, UNICEF, the Ford Foundation, Fulbright, the National Endowment for Democracy (NED), and the National Democratic Institution for International Affairs (NDI) among others, forms a dominant factor in sustaining research activities carried out by both university research centres and private research centres (Al-Sayyed Al-Aswad 2006). Research in these projects tends to be focused on societal issues, particularly those arising as a consequence of globalisation in different Egyptian contexts

– peasants and agricultural laborers now being employed in factories, for instance. Given the research funding shortages briefly sketched above, these projects have become a major opportunity for many university professors, along with employment abroad – particularly in Arab Gulf countries. This sort of "contract research" is not harmful *per se*, of course (Shami 1989); but considering the scarcity of research carried out in university departments, some of which are also doing this kind of research, its share within the knowledge produced in and on Egypt seems to be at odds with the possibility for the emergence of a well-formed, indigenous way to conduct anthropology, particularly from a theoretical point of view.

Parallel to this, there is a surge in anthropological studies conducted in Egypt outside the academy. Compared to the discourse of crisis affecting university departments, the difference could hardly be greater:

> "Anthropology has been "born again" in Egypt. National policy makers and international donors working in Egypt (and perhaps elsewhere) have an increasing awareness of the contribution that anthropology can make to social research and human understanding. In fields as diverse as health sciences and medicine, demography, and other population sciences, ecological and environmental research and advocacy [...] anthropologists who had barely been humoured previously are now sought and heard" (Sholkamy 1999: 119).

But immediately after this statement, the author acknowledges that "on closer inspection one finds that anthropology has been born again as a collection of qualitative methods [...] Quasi-anthropological techniques are in demand, not anthropology with its precepts and concepts" (ibid.).

To return briefly to the 2015 conference, this attitude was again reiterated by the head of the governmental funding agency, when confronted with the critique that there were too few social science projects being funded – his position was that "we" need more social sciences. What he meant was that there was a need for social sciences within scientific projects, in which social scientists' participation is strictly circumscribed to performing specific tasks within projects that are largely, if not completely, already predetermined.

Who counts as an Egyptian anthropologist?

As indicated in the introduction, an Egyptian anthropology cannot be easily separated from the anthropology of Egypt. Even differentiation criteria that look solid from afar, such as national origin or the place where scientific qualifications were obtained, do not stand closer inspection. All the personalities I sketched in this chapter underwent a decisive formation abroad, and this is still a general rule for anthropologists in Egypt; moreover, many of those who receive training at home will probably be employed abroad, at least temporarily, or will take part in research projects designed and funded elsewhere. The degree of mobility, already a rather salient characteristic of scholars worldwide, is, in the case of Egypt, and anthropology in particular, very strong. From the other side of the equation, how long does it take for a non-Egyptian to be considered an Egyptian anthropologist, or at least partaking in Egyptian anthropology? The question is again less clearcut than it may seem, to the point that Nicholas Hopkins in his mapping of the first two thirds of a century of Egyptian anthropology mixes scholars who were Egyptian by nationality with many who weren't, such as Hans Alexander Winkler or Michael Gilsenan, among many others.

Here, in an exploration of decolonisation, I think it's particularly useful to present a discussion on the indigenisation of social sciences, which I think is a misleading way out of the conundrums provided by knowledge and power; even more so, the parallel perspective of an islamisation of anthropology, whose heyday was much shorter. The validity of this perspective has been discussed at length and questioned (for example, Lange 2005). Morsi, Saad, Nelson and Sholkamy (1990), that is three indigenous and one non-indigenous anthropologists who "conducted fieldwork in and on the Arab world for nearly twenty-five years", discuss the issue of how to create an Arab social science. This should rest on some foundations that are culturally specific while being inscribed in the global social sciences, thereby not denying the scientific aspect of the discipline in favour of a specific nationalism. They proposed recognising that its foundation lies in the methodologies used, and not in any theoretical or ideological assumptions – particularly because the "problems of 'our' people cannot be understood without making reference to global situations" (Morsy et al. 1990: 91). These authors clearly identify the "political, epistemological and methodological issues raised in the 'call for the indigenisation of the social sciences'" (ibid.: 102). If an indigenous anthropology is defined as "the practice of social and cultural anthropology in one's own national setting", the concept obscures a number of critical issues in the production of social science knowledge – in particular the idea that nationals studying their own societies, or women studying their own gender, would produce more authentic and relevant knowledge. This aim is a valid and worthy one, but it has to be reached by questioning the research relationships through which anthropological knowledge is produced, namely the positioning of the researcher, wherever she may be, and the peculiar configuration of the power/knowledge nexus that transpires from her research (ibid.: 103).

If anthropology is not the science of the "primitive", but a comparative science of society and culture, the practice of which needs to be assessed from the methodologies employed and the positionality of the researcher, then an Arab or Egyptian anthropology must question received wisdom not on ideological grounds, but on the basis of field research; this research may be conducted by native or non-native anthropologists who actively engage in Arab society and do not consider it simply a source of data (Shami 1989). A special issue of the *Cairo Papers in Social Sciences*, edited by Seteney Shami and Linda Herrera, was entirely devoted to the question of producing social sciences, and anthropology in particular, in and on Egypt; all papers included in the collection explicitly discuss the positionality of the researcher in the research process and in the ethnographic encounter (Shami and Herrera 1999).

Producing anthropological knowledge in Egyptian universities today

The 2011 revolution represented a window of opportunity to bring forth change in the institutional conditions of doing research. I thus decided to focus my research project on the doctoral phase as being particularly significant in the life of the institution, a crucial and formative moment in the production of knowledge, a true moment of passage between different statuses. In short, a liminal moment in the life of the institution, at a time in which change seemed possible. The actual project started in 2014, when the momentum of change was already gone, although that was not immediately apparent at the time. The picture I gathered over the next two years was the one of the institution in crisis, as briefly sketched above; I was repeatedly told that the doctorate is not a formative moment, and generally not to be confused with doing actual research. This assessment is coupled with the informally widespread wisdom that there is no research in the social sciences and humanities at Egyptian public universities, particularly at the doctoral level. For a detailed discussion of the project and of its results I refer to Cantini (2018, 2021), but for the purpose of the discussion here, the main point is that the discourse of crisis was solidly in place, as well as the idea that actual research is produced elsewhere, mostly in consultancy or in institutions abroad. At the same time, however, parallel to the devaluation of local production of knowledge there is a continuous growth in the numbers of people pursuing postgraduate studies, coupled with the continued practice of sending some students abroad on government fellowships, or increasingly on foreign-paid scholarships; for example, over a thousand students went to Europe as part of the Erasmus programme. The discourse of crisis continues to side with an expansion of the system, at least in sheer quantitative terms.

This growth indicates that the need for obtaining research credentials and qualification continues unabated, and that one would be well advised not to take the discourse of crisis at face value, but rather to contextualise it in the broader international knowledge economy and in relation to the new imbalances that are created in the process. The project was not only relevant from its content point of view, but perhaps even more so in its structure as a training process for the scholars involved, all doctoral candidates at Egyptian public universities. It was thus an attempt to set up a truly collaborative project, with the aim of readjusting the imbalance between local and international production of knowledge, and of increasing communication between different contexts.

In order to ascertain what exactly concerns anthropology as an academic discipline, a recent research project surveyed the MA and PhD theses produced at anthropology departments in different Egyptian universities between 2000 and 2016 (Badawi 2017). The article confirms the crisis discourse, and the diminished importance of universities as places of knowledge production. Limiting the analysis to Alexandria University, in 1996–2016 there were 35 MA and 12 PhD theses compared to 47 and 33 respectively in the 1976–1996 period. All over Egypt, in the period 2000–2016 there were 91 MA and 35 PhD theses in anthropology, with Alexandria as the first and Beni Suef as the second most productive departments. The latter was established as a Cairo University branch in the 1970s and became independent in 2006. Other departments that admit theses in anthropology are found at Cairo, Ain Shams, Helwan, Minia, and Sohag (ibid.: 293). More than two-thirds of theses were written by female scholars and almost all of them by Egyptians, with a handful of exceptions in Cairo and Ain Shams. As already observed, there is a notable concentration in the supervision of theses, despite some dominant supervisors being already retired. Although the theses produced have a strong regional focus, the research themes are very scattered (ibid.: 303–304) and lack a proper research agenda or even a strategy.

Doing research in Egypt is also a dangerous endeavour, however, and any discussion of knowledge production cannot ignore the real perils and risks that researchers have to fear. This is not a new phenomenon, sadly; Al-Gohari, an anthropologist mentioned above, insisted that the social sciences were most damaged by the 1952 revolution with the introduction of the requirement to apply for research permission from CAMPAS (Central Agency for Public Mobilization and Statistics) which was run by an officer under the Egyptian military regime. This was the moment that marked the beginning of the "criminalization of collecting data and conducting research" (Abaza 2010: 202). The revolution instituted a law making it illegal to gather information that could harm the state; and the law also put restrictions on fieldwork and the collection of statistics. According to Al-Gohari, after this law was passed in the 1950s it established the association of sociological investigation with spying and information-gathering for the "enemy" (ibid.). Doing research was criminalised, although the actual red lines were never clear, even less made public, making it impossible for researchers to know in advance what could become an issue. Even when research was commissioned by the government, if the results were not deemed acceptable for the public, they would be withheld, along with the research materials collected. Conflicts over the permissible limits of doing research, particularly on ethical themes that are normatively regulated in Islamic jurisprudence, have been present since the beginning of the university in Egypt, where both Egyptians and non-Egyptians were involved. Yet the criminalisation of doing research seems to have reached new heights since the 1990s, perhaps linked with the reforms that were altering the structure of Egyptian economy and society, or with the relative pluralisation of Egyptian society.

In an article aptly titled *Why is Anthropology So Hard in Egypt*, Hania Sholkamy reflected on the case of Saad Eddin Ibrahim, a political scientist who was among the young enthusiasts of the revolution in the 1960s. Ibrahim's work has acquired a considerable political and economic capital since then due to its direct relation with presidents in Egypt and in the USA, and because of his disgrace over allegations of being a spy for foreign agents (for details on his case, see also Abaza 2010). According to Sholkamy, the real issue in this *cause célebre* was the right of the authority to shape and give currency to truth; the Egyptian press demonised the accused and criminalised his whole profession. "They conveyed that it was not only Dr Ibrahim who was guilty of wrong doing, but all those like him who conduct research in towns and villages, defame the national image of their country and attend conferences abroad where they describe and share their research findings.

This sad situation expresses a crisis in the understanding of research and in the proscription of a censorship of its findings" (Sholkamy 1999).

After the 2011 revolution and the moment for change that it offered, the situation changed dramatically, with months and years of clashes on campuses and deep interventions into the structure and the functioning of the university. In the current political climate, it should thus come as no surprise that, when research is discussed, it is the emphasis on the needs of the state which is being put forward again, as in the post-independence period. For example, in November 2018, the Alexandria University's Council for Postgraduate Studies and Research demanded that the dissertations presented to the institution must comply with Egypt's 2030 Vision, a long-term development scheme launched by the president, Abdel Fattah el-Sisi, in 2016.

Decolonising anthropology in Egypt

In this chapter I have tried to provide a succinct account of the development of anthropology in Egypt, as well as on the overall possibilities of doing research, in and on the country. The situation looks as complicated as ever, and most of the problematic issues sketched here are continuing unabated: the crisis in the university as an institution, and consequently the very possibility of the establishment of anthropology as a discipline in its own sense, capable of accumulating knowledge and making theoretical interventions that are relevant outside the country's borders; the understanding of anthropology as being at the service of the developmental needs of the state; and the ongoing repressions that make the pursuit of independent and critical inquiries rather dangerous. Even the few places where anthropology had some chance of establishing itself, such as the AUC, are currently being run according to a neo-conservative model in which the pursuit of knowledge for knowledge's sake does not seem to be a priority, even less so when social sciences are concerned.

Moreover, the global imbalances in the distribution of power and knowledge that subordinate the "Global South" to epistemic necessities set elsewhere were also evident during the 2011 revolution. As soon as Egypt had acquired a new relevance in the eyes of the world, scores of journalists, commentators, and researchers rushed to jump on the bandwagon, struggling to gain privileged access to the protagonists of *Tahrir*, the rebellious yet peaceful and ingenuous youth movement that had been able to inspire the world. In the process, local academics were (again) treated as providers of data and of contacts, as facilitators and fixers, while the real analyses were done elsewhere (Abaza 2011).

Despite all, however, the moment for change offered by the 2011 revolution did touch anthropology as well. For perhaps the first time since its inception, many young Egyptians came to see anthropology not as a primitivist or a colonial science, but rather as a discipline that could offer a methodology closer to the street and to the voice of the people, perhaps even able to speak their language. PhD theses proposing to examine the revolution or addressing topics such as creativity in sociological research were registered, and although the political conditions changed all too soon, the interest in the ethnographic method, and perhaps also in anthropology as a discipline that has the potential to imagine things otherwise, is still there. In recent years, a series of initiatives have tried and are still trying to intercept this interest and to provide young researchers with the theoretical tools to be able to construct their own research topics – institutes such as Cairo Institute of Liberal Arts and Science (CILAS), for instance, the reading group in the theory of anthropology set up at the Bibliotheca Alexandrina in collaboration with the Center for Future Studies, or the *Seket al-Maaref* (Knowledge Rail), an initiative run by some anthropologists to read theoretical texts outside the university campuses.

These are just examples of recent initiatives that are proliferating despite the discouraging political climate, and the more broadly problematic conditions of doing research. They are a sign that the interest in anthropology as a discipline open to experiment and contamination with other experiences, particularly the arts, is gaining recognition in Egypt. Perhaps the growth of anthropological interest in Egypt during recent years within the overall discipline, also linked to the 2011 revolution, has been instrumental in creating more occasions for contact between local and foreign scholars, and funding opportunities to obtain scholarships abroad; collaborative projects have probably played a role too. In this sense, one could conclude that, for those who care to watch closely – and despite all constraints – anthropology in Egypt is already becoming decolonised.

Bibliography

Abaza, M. (2010): "Social Sciences in Egypt: The Swinging Pendulum between Commodification and Criminalization." In: M. Burawoy (ed.), Facing an

Unequal World: Challenges from Sociology, Sydney: International Association of Sociology, pp. 187–212.

Abaza, M. (2011): "Academic Tourists Sight-seeing the Arab Spring." In: Ahram OnLine, September 26, 2011.

Ahmad, A. (1953): Qâmûs al-ʿâdât wa-l-taqâlîd wa-l-taʿâbîr al-miṣriyya [Dictionary of Egyptian Customs, Traditions, and Expressions] (al-Qâhira [Cairo]: Lajnat al-taʾlîf wa-l-tarjama wa-lnashr).

Al-Sayyed Al-Aswad, (2006): "Applied Anthropology in Egypt: Practicing Anthropology within Local and Global Contexts." In: NAPA Bulletin 25, pp. 35–51. https://doi.org/10.1525/napa.2006.25.1.035.

Assal, M. (2018): "Anthropology in Sudan." In: The International Encyclopedia of Anthropology, John Wiley and Sons, pp. 1–13. https://doi.org/10.1002/9781118924396.wbiea2101.

Awad, L. (1963): al-Jami'a wa-l-mujtam'a al-jadida [The University and the New Society], Cairo: Dar al-Qawmiyya lil-Taba' wa al-Nushr.

Badawi, A. M. (2017): "Research Priorities of the Egyptian Anthropological Community (in Arabic)." In: Majalla al-Taghir al-Ijtima´i wa al-´Ilaqat al-´Ama 3, pp. 285–314.

Boëtsch, G. (1995): "Noirs ou blancs: une histoire de l'anthropologie biologique de l'Égypte." In: Egypte/Monde Arabe 24, pp. 113–138. https://doi.org/10.4000/ema.643

Cantini, D. (ed.) (2016): Rethinking Private Higher Education: Ethnographic Perspectives, Leiden: Brill. https://doi.org/10.1163/9789004291508

Cantini, D. (ed.) (2018): "al-Intag al-Ma'rifa fi al-Jami'at al-Masriyya [Knowledge Production at Egyptian Universities]." In: Idafat: al-majallah al-arabiyya li-l-Ulum al-Ijtima'iyya [Idafat: the Arab Journal of Sociology] (special issue) 41–42, pp. 13–150.

Cantini, D. (ed.) (2021): Bounded Knowledge: Doctoral Studies in Egypt, Cairo/New York: American University in Cairo Press. https://doi.org/10.2307/j.ctv2ks6x56.

Dessouqi, N. (2021): "Doctorate Studies at the Sociology Department, Cairo University: A Case Study." In: D. Cantini (ed.), Bounded Knowledge: Doctoral Studies in Egypt, Cairo/New York: American University in Cairo Press, pp. 51–72. https://doi.org/10.2307/j.ctv2ks6x56.7

El Shakry, O. (2007): The Great Social Laboratory: Subjects of Knowledge in Colonial and Postcolonial Egypt, Stanford, CA: Stanford University Press. https://doi.org/10.11126/stanford/9780804755672.001.0001.

Ferrié, J.-N. (ed.) (1995): "Anthropologies de l'Égypte 1." Egypte/Monde Arabe 24. https://doi.org/10.4000/ema.619.

Hamdan, J. M. (1970): Shakhsiyyat Misr [The personality of Egypt], Cairo: Alam al-Kutub.

Hijazi, M. I. (1986): "Al-azma al-rahina li-'ilm al-ijtima' fil watan al-'arabi" [The current crisis of sociology in the Arab world]. In: Nahw 'ilm ijtima arabi [Toward an Arab Social Science] Beirut: Markaz dirasat al-wihdah al-'arabiyah, pp. 13–44.

Hopkins, N. (2010): "Anthropology in Egypt 1900–67: Culture, Function, and Reform." In: Cairo Papers in Social Science 33/2.

Hopkins, N. (2014): "Mohamed Galal (1906–1943): A Pioneering Egyptian Anthropologist." In: François Pouillon/Jean Claude Vatin (eds.). After Orientalism: Critical Perspectives on Western Agency and Eastern Re-Appropriations, Brill. https://doi.org/10.1163/9789004282537_014.

Hopkins, N./Mehanna, S. (2011): Nubian Encounters: The Story of the Nubian Ethnological Survey 1961–1964, Cairo/New York: American University in Cairo Press. https://doi.org/10.5743/cairo/9789774164019.001.0001.

Lange, K. (2005): Zurückholen, was uns gehört. Indigenisierungstendenzen in der arabischen Ethnologie, Bielefeld: transcript Verlag. https://doi.org/10.1515/9783839402177.

Morsy, S./Nelson, C./Saad, R./Sholkamy, H. (1990): "Anthropology and the Call for Indigenization of Social Science in the Arab World." In: E. Sullivan/J. Ismael (ed.), The Study of the Contemporary Arab World, University of Alberta Press.

Najjar, F. (1976): "State and University in Egypt during the Period of Socialist Transformation, 1961–1967". In: Review of Politics 38/1, pp. 57–87. https://doi.org/10.1017/S0034670500015473.

Perrin, E. (2002): "Le creuset et l'orfèvre: le parcours d'Ahmad Amîn (1886–1954)." In: Revue des mondes musulmans et de la Méditerranée, 95–98, pp. 307–335. https://doi.org/10.4000/remmm.238.

Perrin, E. (2005): "Le musée d'Ethnographie de la Société de Géographie d'Égypte. " In: Gradhiva: revue d'histoire et d'archives de l'anthropologie, Musée du quai Branly, pp. 5–29. https://doi.org/10.4000/gradhiva.455.

Reid, D. M. (1990): Cairo University and the Making of Modern Egypt, Cambridge: Cambridge University Press. https://doi.org/10.1017/CBO9780511599651

Reid, D. M. (1993): "The Egyptian Geographical Society: From Foreign Laymen's Society to Indigenous Professional Association." In: Poetics Today 14/3, pp. 539–572. https://doi.org/10.2307/1773284.

Roussillon, A. (1985): "Sciences sociales et réformisme. Une sociologie de l'Égypte éternelle: Postface." In: L'Histoire que je porte sur mon dos: Mémoires [online]. Le Caire: CEDEJ – Égypte/Soudan.

Roussillon, A. (2002): "Sociologie et identité en Égypte et au Maroc. Le travail de deuil de la colonisation." In: Revue d'histoire des sciences humaines 7, pp. 193–221. https://doi.org/10.3917/rhsh.007.0193.

Shahshahani, S. (1986): "History of Anthropology in Iran." In: Iranian Studies 19/1, pp. 65–86. https://doi.org/10.1080/00210868608701669.

Shami, S./Herrera, L. (eds.) (1999): Between Field and Text: Cairo Papers in the Social Sciences, Cairo: American University in Cairo Press.

Shami, S. (1989): "Socio-cultural Anthropology in Arab Universities". In: Current Anthropology 30/5, pp. 649–654. https://doi.org/10.1086/203797.

Sholkamy, H. (1999): "Why is Anthropology So Hard in Egypt?" In: S. Shami/L. Herrera (eds.), Between Field and Text: Emerging Voices in Egyptian Social Science, Cairo Papers in Social Sciences 22/2, pp. 119–138.

Zayed, A. (1995): "Seventy Years of Sociology in Egypt". In: Cairo Papers in Social Sciences 18/3, pp. 41–71.

Thoughts on "Coloniality" and "Africanity"

Scholarship in African Universities and African Studies in Germany

Abimbola O. Adesoji and Hans Peter Hahn

Introduction

Ideally, scholarship should be universal, without any boundary or limitation: it should not be constrained by such extraneous factors as race, ethnicity, religion, class, ideology or worldview. But the reality is that most of the humanities disciplines that emerged or grew in the 19th century were defined in an intellectual environment that was heavily influenced by Eurocentrism and colonialism.[1] Subsequently coloniality persisted, particularly with regard to the colonisation of ideas and minds despite efforts at decolonisation. The wider implication of this was the emergence of dominant voices or perspectives that dictated the direction of scholarship worldwide. Although there have been increasing calls globally for the decolonisation of knowledge during the last forty or fifty years, particularly within the Euro-American context, attempts at decolonising knowledge in African universities more than six decades after political independence have been ineffectual. Relatively recent protests by students in the Southern African region reflect the frustrations experienced by African students and their determination to give expression to it.

It is worth pointing out that in Euro-American circles, the colonial character of the humanities[2] – as well as the sciences – has either been unacknowl-

1 This is true of many other disciplines that emerged and developed during the 19th century. Even long-established disciplines, such as medicine, have changed their scope and range of methods considerably. Robert Koch's studies of infectious diseases provide an example of this (Cf. Bruchhausen 2019).

2 We distinguish here between the "Eurocentric" and the often implicit "colonial" ideology. In Eurocentricism information is collected about issues regarding European so-

edged by representatives of the disciplines concerned or denigrated as being irrelevant. Calls to consider the colonial character of scholarly knowledge about history and society began in the 1970s (Asad 1973; Leclerc 1972), but, at that point reflections on implicit biases were often marginalised or dismissed as irrelevant to the questions perceived as central at the time. The colonial legacy of knowledge in many disciplines, including anthropology, history, archaeology and philosophy, was then perceived as a phenomenon of the past. Leading representatives of these subjects denied colonialism's lasting influence on the production of knowledge prior to, and even after 1970. The conviction of their own scientific objectivity dominated.

Although efforts at promoting knowledge about Africa and African studies have advanced, they have nevertheless remained largely situated within the Western epistemological and ontological paradigm. Consequently, their methodological and theoretical approaches and perspectives have been largely informed by Western thinking.

One indicator of the neglect of non-Western perspectives on knowledge and thought is the marginal status held by African philosophy until very recently. It was not until 2015, for example, that the first overview of African political philosophy was published in German (Dübgen and Skupien 2015). Another significant example is the evaluation provided by Kwame Nkrumah whose recognition as a philosopher was delayed until 2020.[3] In particular, the logic of catch-up development, which was dominant in Africa, was based on the idea that fundamental ways of understanding and evaluating society in Africa did not require independent reflection. Science in Africa was seen primarily as a matter of applying models previously developed by, in, and for Europe. This attitude also affected the structure of universities in Africa:

cieties, cultures and history and thus implies an (often unconscious) neglect of other parts of the world. "Colonial ideology" is linked, but, in addition, refers to non-European societies and cultures as "inferior". Very often, images of the supposedly inferior character appear without further explanation, and the legitimising character of such images is not made explicit. Locating the "other" within the past, or defining them as being "child-like", or as "people without history", is undeniably related to a colonial ideology (Blaut 1993; Quijano 2000).

3 As has been shown only recently, Nkrumah's work on "consciencism" can be read as a critique of Immanuel Kant's work (Nkrumah 1964). However, European scholars failed to acknowledge this because a dialogue involving Kant versus Nkrumah was not considered plausible (Uimonen 2020).

"applied sciences" appeared more important than subjects that did not seem to have any direct applications.

As these examples show, it is important to improve global awareness about Africa and to ask pertinent questions: for example, how is Africa being studied? In other words, how is the knowledge about Africa that is discovered by Africans and non-Africans alike being made known? Who determines what is to be studied or known about Africa? What perspectives, approaches or worldviews influence knowledge production both generally and concerning Africa specifically?

Of particular importance to African studies are the African Studies associations situated in Western universities and countries (e.g., African Studies Association in the UK (ASA-UK), German African Studies Association (VAD), Associazione per gli Studi Africani (Italy), Nederlandse Verengiging voor Afrikastudies, and others).[4] The emergence, programmes and activities of these organisations, spanning several decades, have promoted the field of African studies, but, in a way, they have also constrained it. Critical questions still need to be asked: how inclusive, involving or accommodating are Africans in these associations or in other fora in their quest for knowledge production both in Africa and about Africa? The way in which the issues involved are responded to has raised fundamental questions about the nature of scholarship in and about Africa.[5]

Coloniality, decoloniality and Africanity: A discourse

As a tenacious project, colonialism has reinvented itself in different forms including neocolonialism and coloniality. Such reinvention is testimony to the fact that the process of decolonisation that led to the independence of most African states did not inevitably bring about the end of colonialism. Rather, the independence period marked the beginning of another long phase in which the hegemonic structure of asymmetric forms of relations between the Western

4 At the ECAS conference in Edinburgh (12th June 2019) a panel organised by one of the authors (H.P. Hahn) brought together leading representatives of the African Studies Association mentioned above.

5 It is recognised today that in the case of the US-American ASA, scholars were not quite inclusive. In contrast, white domination and the exclusion of black scholars was initially an obvious, although never made explicit, strategy (Allman 2019).

world (Europe and North America) and the so called "Third world" prevailed. After referring to Ngũgĩ wa Thiong'o's work (1986), and summarising his own position, Ndlovu-Gatsheni (2015a: 488) provides the following definition:

> "Coloniality is, therefore, an invisible power structure, an epochal condition, and epistemological design, which lies at the center of the present Euro-North American-centric modern world. At the center of coloniality is race as an organizing principle that not only hierarchized human beings according to racial ontological densities but also sustains asymmetrical global power relations and a singular Euro-North American-centric epistemology that claims to be universal, disembodied, truthful, secular, and scientific"

Africa falls within what scholars have described as part of the postcolonial neocolonised world and the postcolonial neocolonial world (see Comaroff and Comaroff 1993; Geschiere 1995; Ndlovu-Gatsheni 2013). While the former term describes the structural, systemic, cultural, discursive, and epistemological pattern of domination and exploitation of Africa by the colonisers, the latter not only captures failed liberation projects that gave birth to independent-dependent African states, but also depicts an "African state of 'becoming' that never materialized" (Ndlovu-Gatsheni 2013: 4). Characteristic of coloniality are the Western projects of modernity, globalisation, entrenched epistemological hegemony, subtle manipulation and perennial domination of the postcolonial neocolonised economy and government. As argued by Ndlovu-Gatsheni, at the core of struggles in the postcolonial neocolonised world is the question of knowledge production and dissemination, besides the important issue of freedom (ibid.: 19–28). Coloniality and decoloniality, therefore, are not subjects locked in the past, but they simultaneously represent a memory of the past that either validates or establishes limitations on the valuation of the present, and promotes a movement towards the emancipation of postcolonial decolonised people.

Still following Ndlovu-Gatsheni (2015b: 15), coloniality should be distinguished from colonialism as it refers to "long-standing patterns of power" and is relevant well beyond the activities of various colonial administrations in the past. This argument is further explored by Maldonado-Torres (2016: 10), who aptly captures the core of coloniality and decoloniality (as distinct from colonialism and decolonisation) when he notes that "coloniality and decoloniality refer to the logic, metaphysics, ontology, and matrix of power created by the massive processes of colonization and decolonization". The enduring nature of this matrix of power imprinted by the Euro-North and America gave birth to the hegemonic universalisms known as "Western civilisation" and "Western modernity."

Hence, if coloniality reveals continuous dehumanisation of the world and perpetuates Western hegemony, decoloniality represents the drive to rehumanise the world. It is a movement that seeks to break "hierarchies of difference that dehumanize subjects and communities and that destroy nature, and to the production of counter-discourses, counter-knowledges, counter-creative acts, and counter-practices that seek to dismantle coloniality and to open up multiple other forms of being in the world." (Maldonado-Torres 2016: 11).

Decoloniality concerns both economics and academia. While the focus in economics is on overcoming unequal "terms of trade", the idea of a paradigm shift applies to scholarship (Misgav 2010). Decolonial science, whether history or anthropology, must develop a different conception of Europe (Chakrabarty 2000). Overcoming coloniality has so far been only partially successful. Despite the rich forms of institutional education present in pre-colonial Africa, European colonisers succeeded in imposing the phenomenon of under-education, along with exploitative research practices and racist theories of knowledge, using these as instruments of colonial subjugation (Kessi et al. 2021). Scholarly education and research were never intended for all thinkers and knowers and not meant to promote inclusive human progress. Accordingly, the Western university and intellectual property rules first achieved their diffusion through global imperialism and the global knowledge economy that prevails to this day. An indication of this thesis is the dominance of English as the language of science, a usage which suppresses indigenous languages and regionally specific scientific structures. Language, knowledge and epistemic orders reinforce each other.[6]

Education and knowledge production in postcolonial Africa is characterised by independent-dependence. African scholars and students are

6 This holds true in spite of the fact that English is currently becoming a *lingua franca* for scholarly texts. Furthermore, it should be highlighted that monolingualism in science exerts considerable negative effects on scholarship and the intellectual potential of authors (cf. Horner et al. 2011).

trapped in the Western epistemological model.[7] This independent-dependence – understood as African peoples' perennial attachment to Euro-American thoughts, knowledge production, and dissemination – has been interpreted by scholars as "coloniality of knowledge" (Quijano 2000). Maldonado-Torres (2016) notes that apart from national culture and the media, education and academic scholarship are potent instruments for reproducing Western-centric modernity and colonialisation in the neocolonial colonised world.

Knowledge production, dissemination and circulation are important ways in which the Western hegemony has been entrenched and maintained since the colonial period. Coloniality of knowledge is characterised by the tendency to marginalise non-Western scholars and epistemologically designed to keep them on the lowest rung of the knowledge ladder. This is sometimes perceived as a war of attrition by scholars from the Global South. The enduring psychological and epistemological effects of coloniality are best termed by Ndlovu-Gatsheni in the citation above.

The coloniality of knowledge is maintained by the Euro-North American states through the use of a number of instruments. The first is the remapping of the world and an associated racial hierarchisation of knowledge. Ex-colonies are remapped and renamed based on new racial categorisations and assumed intelligence: for example, the Middle East, Asia, Latin America, the Caribbean and Africa.[8] Second, coloniality is maintained through appropriation and monopolisation of all instruments of knowledge production and dissemination. From the education curriculum through to the method of knowledge production and subsequent publication, Euro-North American states and teaching structures dominate.

The third instrument is the deployment of subtle seduction via the affordances of funding and the media. The politics of academic funding has contributed to subjective knowledge production and dissemination in the Third World, a development aptly captured by the saying, "he who pays the piper calls

7 Regardless of whether one considers Western epistemologies to be multiple or homogenous, it is obvious that differences on a global scale are more significant. Anthropological theory on ontologies was, amongst other things, inspired by the fundamental differences between Amazonian and Western ontologies (cf. Descola 2005; Holbraad and Pedersen 2017).

8 Alternatively, it would have been very possible to create a cultural space like the "Indian Ocean littoral", or to base educational institutions on trans-Atlantic relations.

the tune" as noted by Ndlovu-Gatsheni (2015a: 485–486). These instruments are still used to perpetuate the coloniality of knowledge in the postcolonial colonised world.[9] Current principles of international cooperation and development practices do not always sufficiently reflect the continuity of marginalisation and poverty. Thus, financial support is often designated for predefined purposes and limited to offering short-term assistance simply to achieve predetermined objectives, whereas the development of local academic knowledge and proper reflection on the nature of development is not supported (Hahn 2021).

The title of this contribution uses the term "Africanity" in order to relate to efforts directed towards making African perspectives known and heard, but without necessarily eliminating informed scientific knowledge. Africanity connotes the idea of academic freedom, otherwise conceived as Africanisation or de-corporatisation (Nyamnjoh 2015). Nyamnjoh identifies insufficient will and an absence of sustained commitment by African scholars – rhetoric apart – as some of the factors responsible for the non-Africanisation of curricula, pedagogical structures, or epistemologies in African universities. Despite the fact that universities have Africanised their personnel, revisions have not been effected in a systematic and productive manner following decolonisation (Nyamnjoh 2015; Ndlovu-Gatsheni 2018). Drawing from the insight that identities are complex and often composite, and that there is a lot more, and also a lot less, to things than meets the eye, Nyamnjoh argues that it is not merely because one *is* or *appears* African that one is necessarily going to be critical of colonial intellectual traditions, rituals, and habitus in one's teaching and research, or offer a menu sensitive to local realities and "endogenous" epistemologies (Nyamnjoh 2015; Hountondji 1997).

Nyamnjoh therefore argues for epistemological inclusivity at African universities, a goal to be achieved by going outside the academy and drawing inspiration from the personal stories and creative imagination of popular Africa. Such sources were ignored under colonialism and regarded as being too "savage" and "primitive" to share a table with European colonial enlightenment. Consequently, they were often misrepresented in the postcolonial era not only by ill-adapted curricula, epistemologies, and theories, but also by many academics and scholars who subscribed to trans-Atlantic scholarly

9 Some scholars are therefore calling for a fundamental change, in order to create the conditions for an alternative academic structure (cf. Marboeuf and Ben-Yakoub 2019).

canons, practices, and standards of value in knowledge production and consumption (Nyamnjoh 2015; Mamdani 2007). Nyamnjoh also makes a case for intentionality in critically questioning (deconstructing) conventional theories, methodologies, and research.

Such deconstruction of existing concepts, rules of engagement, procedures, and processes of scholarship would contribute to promoting African universities and the research networks they fund. Importantly, it would also forge awareness of the risks of intellectual bandwagonism that may result when Africans unquestioningly participate in research and debates on themes already determined and conceptualised by others. This matters when those "others" operate outside African social, cultural, political, historical, and geographical contexts and experiences. Often there is little problematisation of the frameworks of the theories and methodologies at play. Such bandwagonism, shaped by intellectual fashion designers, with little or no regard for the African contributor or consumer – with provision mainly for lecturers and students reduced to potted plants and clearing officers for cheap and untested and often ill-adapted intellectual and academic imports – is a persistent serious threat to Africa's intellectual affirmation (Nyamnjoh 2012: 138).

An impressive example of the possibilities created by a change of perspective is constituted by a range of disparate works on the history of the former German and French colony of Togo. For forty years, the work of a former French administrator, Robert Cornevin (1959), was considered the standard work on the history of Togo due to its comprehensive consideration of both German and French sources. From 1997 onwards the former president of the University of Lomé, Nicoué Gayibor, set out to provide a "History of the Togolese" instead of a "History of Togo" (Gayibor 1997). With this nuance of the title, Gayibor refers to the colonial character of the country's territorial borders, and at the same time focuses on the different population groups, religions, occupational and other social groups. Instead of presenting a grand synthesis (as did Cornvin), Gayibor assembled contributions from numerous Togolese and international authors. Although this work is not free of controversial and possibly subjective opinions of individual authors, the reader can clearly discern an attempt to shake off colonial historiography and to unfold the possibilities of an independent decolonial perspective on the history of the people of Togo (Glasman 2008).

In numerous departments of history in many universities in African countries, historians have set out to revise the historiography hitherto dominated by Western authors. Joseph Ki-Zerbo should be mentioned here as one of the outstanding protagonists. The appropriation of African historiography makes a valuable contribution to the articulation of regional and national identity and such efforts can be understood as an important step towards decolonisation. Similarly, Nyamnjoh argues for making explicit the elements of knowledge that have hitherto been suppressed or largely ignored. His plea is for new approaches that respond to contemporary challenges and aspirations (Nyamnjoh 2012). To some extent, historians have managed to do this without resorting to the loud but stark reminder of South African scholars such as Ndlovu-Gatsheni or Maldonado-Torres.

Still following Nyamnjoh, the Council for the Development of Social Science Research in Africa (CODESRIA) might be an institution capable of uniting such academic efforts on a level that goes beyond individual universities and countries in Africa. CODESRIA empowers scholars to engage in robust academic debates and academic freedom programmes across campuses from Dakar to Dar es Salam, Ibadan and Kampala, as well as enabling them to publish in Africa (Diouf and Mamdani 1994; Mkandawire 2005). Of importance here is the fact that some decolonisation activities have been in progress for at least 30 years despite the devastating picture sketched in some recent and frequently quoted oeuvres on "coloniality". The relevance of this term relates to the persistence of many of those issues that decolonisation was meant to correct or address. It is the enduring nature of the problem that legitimises such a concept as coloniality.

Understanding and appraising *Vereinigung für Afrikawissenschaften in Deutschland* (VAD) (African Studies Association Germany)

But what about scholarship on Africa in Europe? The following section will take a closer look at Africanist scholarship in Germany as one of this chapter's authors (Hahn) was head of the professional association of African Studies in Germany (VAD) from 2018–2021. This institution has a clear commitment to the decolonisation of academia (with a focus on humanities) and works to advance the public image of Africa in Germany. However, it is appropriate to ask whether the actions of the VAD correspond to its self-image and whether the achievements of such a professional association make a significant contribution to promoting a sustainable and appropriate image of Africa to the European public.

The history of the African Studies Association in Germany followed the trajectory of the growth of African Studies in Germany. It is generally agreed that 1969 was the year in which that association was formally instituted by members of the University of Marburg in Germany, although the development leading to this event had begun long before in the closing decades of the 19th century. Initially known as the Association of Africanist Scholars in Germany, and modelled on the German Oriental Society,[10] this association was originally conceptualised as a regionally defined scientific and interdisciplinary forum. Its objectives were the promotion and exchange of ideas between scholars of African Studies working within the university setting and beyond, surpassing national and disciplinary borders.[11]

Significantly, the history of Africa-related research transcends the beginning of its formal institutionalisation having been initially conceived and dominated mainly by linguists as a forum for new ideas and new interdisciplinary work. The focus later shifted in favour of members coming from the social sciences. Thus, starting from the second half of the 19th century, with the contributions of personalities like Heinrich Barth and Friedrich Ratzel, both geographers, African studies took a leap forward. Further impetus was provided by the involvement of others, such as Leo Frobenius, with his emphasis on the study of "cultural provinces" – a sphere or subject built upon by Bernhard Ankermann and Wilhelm Graebner (on the futile character of this approach to African history see Hahn, 2001). Others, including Richard Lepsius, Carl Meinhof and Diedrich Westermann, built further on the foundation laid so that by the end of the colonial period, African studies had developed systematically as an academic discipline (Probst 2005). Interestingly the aspect of "applicability", in the sense of "serving the colonial idea", that had been explicit from the very beginning of the departments became less pronounced later on but was still an implicit option for defining the presumed utility of these academic institutions. The continuity from "educating colonial administrators" to "educating development experts" is obvious, although this was

10 The famous *Deutsche Morgenländische Gesellschaft*, founded in 1845, claims to be the oldest Orientalist association in Germany. It explicitly includes Africa within its regional scope (https://www.dmg-web.de/page/home_de)

11 Felix Brahm, "40 Jahre Vereinigung für Afrikawissenschaften in Deutschland (VAD), 1969–2009" http://www.vad-ev.de/fileadmin/user_upload/pdf/FelixBrahm-40JahreVAD.pdf; African Studies Association, Germany VAD e.V http://vad-ev.de/en/

reflected only by very few Africanist scholars. An example of such a reflection is provided by Helmut Straube (1971).

Meanwhile, the traditional centres of African studies in Germany, namely, linguistics and anthropology, did not form a joint association until the Marburg initiative which was championed by a group of young linguists. This move was meant to overcome African studies' concentration on linguistics and to widen the term *Afrikanistik* with a view to opening the field for representatives of other disciplines dealing with Africa. The constitution of the Association stipulated, accordingly, that *Afrikanistik* should be conceived as a contemporary interdisciplinary, critical, and self-reflexive project actively engaged in collaboration with African colleagues (Straube 1971: 415–416). Interestingly, with the perception that the original aim of the association – which was a genuine discussion of linguistic questions in an open, constructive dialogue with other disciplines – had failed, the founders of the VAD formed their own conference platform, the *Afrikanistentag*, in Cologne in 1978 (VAD 1970). But the continued growth of African studies in Germany from the 1970s, particularly following the collapse of the Berlin Wall, greatly aided and sustained the association (Probst 2005).

The formal inception of the VAD in 1969 provided ample opportunities for a proper coordination of activities related to the promotion of African studies. Such activities include the promotion of research on contemporary societies in Africa, regular meetings of Africanist scholars to facilitate exchange of ideas and information sharing, especially concerning on-going and proposed studies, as well as grants available. Also included are regular visits to Africa for updating individual and group researchers' knowledge: such visits often being sponsored by the German research foundation, *Deutsche Forschungsgemeinschaft*. Beyond the establishment of study centres and the implementation of programmes (some of which preceded the formation of the VAD) one major method employed by the VAD to promote African studies was the holding or organisation of biannual conferences with specific themes. These events have been directed towards probing specific issues and problems and aimed to generate knowledge rooted in clear academic and practical relevance to the African continent. Some of the conferences have also been geared towards deepening African studies in Germany (Probst 2005: 416).

What is obvious in the promotion of African studies generally in Germany, as in other locations outside of Africa, is the desire to know more about the continent: its land, people and culture, and in the process to deepen German society's understanding of Africa. The need to know provides a basis for a deeper investigation, sometimes necessitating collaborative research spanning both the European and African continents.[12]

Although one might have expected that such collaborative research might provide a basis for efficient and sustainable policy intervention in African affairs, this was hardly ever the case. Instead, it proved impossible to establish routines of policy consultation because political representatives preferred to consult economists, lawyers, or regional experts from private research institutes rather than the VAD. Possibly, the government-critical positions of certain members of the VAD, who often publish in the media, have been sufficient reason for the representatives of various ministries of the federal government not to contact this professional association or to ask for its advice. An open confrontation took place in the 1980s in the context of the anti-apartheid movement, which manifested itself not only in Germany, but in many European countries.

A memorandum, co-sponsored by the VAD and published in 1986, caused considerable criticism of the government's political position on apartheid (Bley 1986). In retrospect, there can be no doubt that this initiative successfully undermined the South African apartheid regime's credibility in Germany, and as a result, state and economic actors gradually withdrew from their involvement there (Kössler 2015). In retrospect, it is relevant to ask whether this influence and the sharp, publicly aired controversy could be described as a "decolonial" activity. Should the VAD have renounced this initiative because of its self-image, or should it have shaped it differently? The answer is not obvious and it is also not easy to decide in which situations precisely solidarity can be considered a decolonial activity (Kössler and Melber 2002).

Further examples of such critical engagement by VAD members could be cited at this point, because the activities undertaken by them are important in shaping the identity of this professional association and stand as visible signs

---

12 It would be a timely and valuable research project to compare the activities of such professional associations throughout Europe. The first initiative of this kind was undertaken by one of this chapter's authors (Hahn) by organising a roundtable at the ECAS 2019 conference in Edinburgh (https://ecasconference.org/2019/panels#7580, last accessed 2022-02-05). Representatives of six national African Studies associations (from Britain, Netherlands, Germany, Switzerland, Italy, and Czechia) participated. The differences in the respective agendas and priorities of these participants proved to be much more substantial than expected. There was considerable disagreement especially with regard to the impact of experts' advice on respective national policies.

of its political positioning in the complex and contradictory field of the German government's Africa policy. Although, undeniably, one can speak about "agency" in this context, it is also obvious that the professional association as a whole has limited possibilities for action. In the majority of cases, the VAD has not succeeded in convincing politicians to act in order to overcome inequality or injustice.

A major failing is the VAD's neglect of Africans and People of Colour (PoC) in Germany. It is a lamentable shortcoming that African European Studies have, so far, not been established in Germany (Espinoza 2019). The VAD should probably have been active in the field over a much longer timespan. As things stand, this professional association's contribution in the fight against racism in Europe is far less than could have been expected (Essed 2019).

Ultimately, history will decide whether such initiatives can be considered useful in terms of "decolonisation", or whether they were futile, and purely academic exercises. It is obvious that the VAD has exerted a limited influence on perceptions and interpretations within the larger context of what has been identified as "coloniality" in the Euro-American public sphere. Pertinent in this regard is a common saying used by historians: "facts are sacred, opinions are free". Whereas facts about African politics as well as facts about history written by African historians cannot vary, different perceptions – often influenced by certain extraneous considerations – play a major role in how those facts are interpreted, the meaning that is made out of them, and the use to which they are put.

A challenge remains for African and German scholars and researchers. The immediate consequence of this is often the sheer impossibility of presenting a balanced perspective on issues. Given the dominance of Western interpretations and perspectives, the permanent suspicion about whether unbalanced or jaundiced perspectives are at play too often results in the relegation of Africa – and the African voice not being heard on African matters.

This problem aligns with Nyamnjoh's observation that "knowledge production takes place in a world of interconnecting global and local hierarchies informed, among other things, by race, place, culture, class, gender, and age" (Nyamnjoh 2019). Given these limiting factors of objectivity and the particular disadvantageness of Africanist scholarship in Europe, a cautious evaluation should be adopted.

In the words of Nyamnjoh, "incompleteness" is a social reality and forms of knowledge generation depend on interconnections, relatedness, open-endedness, and multiplicities (ibid.). Although overcoming coloniality cannot be achieved in an absolute form, such an endeavour will always remain an ideal to follow, despite the fact that it could prove difficult to achieve in day-to-day interactions.

Nyamnjoh argues that the many challenges to academic freedom are shaped in part by identity politics and the hierarchies of humanity, as well as by the racial, ethnic, class, gender, generational, or religious backgrounds of those seeking, denying, or being ambivalent about seeking such freedom – yet he makes a case for "convivial scholarship". This he describes as a scholarship that, rather than dwelling on zero-sum games of absolute winners and losers, instead encourages a disposition of incompleteness and the humility of doubt, finding strength in the themes of interconnection, interdependences, compositeness, and incompleteness of being (Nyamnjoh 2017).

It should be pointed out, however, that although this development could be seen as a relevant issue at the level of the VAD, it is more appropriately situated at the level of universities and centres where teaching and research is done and from where VAD membership is drawn. Arguably this could never have become an issue at the association level but for the existence of the teaching and research centres which define and implement the research agenda and give it power.

Conclusion

Whereas African studies scholarship could perhaps not have attained the level it has reached in the West, scholars are nevertheless persevering to promote studies about Africa. Although African studies associations like the VAD, which have supported studies about and in Africa, have made important advances, they have had to struggle with particular constraints given the limiting influence of the political environment in which they operate.

A remarkable parallel should be pointed out here. As shown with the example of historians in Africa, academic initiatives in Africa have to battle to appropriate the production of knowledge and the formal design of knowledge transfer with societies in Africa, and thus to highlight the specificity of knowledge production in Africa. Similarly, a professional association such as the VAD is repeatedly faced with the challenge of positioning itself critically in a public sphere that often has little interest in the special insights and the state of knowledge of Africanist scholars. In both cases, there are specific concerns that are of essential importance for Africanity, and which have a hard time in an environment of a still dominant coloniality.

As things stand today, it is not easy to decide whether the initiatives of the scholars in Europe and in Africa, mentioned in this article as examples, can be evaluated as part of the struggle against coloniality. It is probably too early to evaluate the contributions of specific initiatives in this struggle. It is also a matter of scholarly honesty, and of respect for future historical assessments, that such evaluations must be left to the historians of future generations.

Respect for the facts, and caution with regard to evaluation and interpretation, should at the same time be taken as a criticism of the South African authors cited several times in this chapter. It is easy to develop a universal paradigm like "coloniality", but it takes a lot of careful empirical work to identify hard facts that allow a reliable assessment of individual events within the framework of this universal paradigm. The progress of knowledge would not be served well if it remained under the blanket judgement of a "coloniality" that has been widespread worldwide for centuries. It is possible to consider what scholars such Sabelo Ndlovu-Gatsheni, Francis Nyamnjoh, and Nelson Maldonado-Torres are doing as a continuation of what Joseph Ki-Zerbo, Nicoué Gayibor, Jacob Ade-Ajayi and others started in their generation. Perhaps what has changed is the increased vocality and the daring attitude of new generations of scholars given the persistence of the same problem and its transmutation (colonialism has transmuted to neocolonialism and is persisting with coloniality). For the sake of historical honesty, it is necessary today to judge the actions of courageous scholars – such as the historians in West Africa mentioned above, or the Africanist scholars in the VAD in their fight against apartheid – with caution and differentiation.

Regardless of what assessments may be made in the future, African scholars in Europe as much as in Africa should work tirelessly, now, not only to sensitise the public to the continuing contradictions of coloniality, but also to promote scholarship on Africa. Indeed, rather than keying to already established or popular buzzwords, or just tagging along, Africans can actually set and drive research agenda both at home and on the international platform provided by various African Studies associations. What is required is proactive commitment and a desire to excel limitations – plus the aid of adequate research funding from governments and organisations on both continents: Africa and Europe.

Bibliography

Allman, J. (2019): "#HerskovitsMustFall? A Meditation on Whiteness, African Studies, and the Unfinished Business of 1968." In: African Studies Review 62/3, pp. 6–39. https://doi.org/10.1017/asr.2019.40.

Asad, T. (1973): Anthropology and the Colonial Encounter, London: Ithaca.

Blaut, J. M. (1993): The Colonizer's Model of the World: Geographical Diffusionism and Eurocentric History, New York: Guilford.

Bley, H. (1986): Südafrika zum Frieden zwingen. Eine Denkschrift, Hannover: VAD.

Bruchhausen, W. (2019) : " Global Health in the Colonial Era: The Expansion of European Medicine." In: European History Online (EGO), Mainz: Leibniz Institute of European History (IEG), 2019, [2021–12-26] .

Chakrabarty, D. (2000): Provincializing Europe: Postcolonial Thought and Historical Difference, Princeton, NJ: Princeton University.

Comaroff, J./ Comaroff, J. L. (ed.) (1993): Modernity and its Malcontents. Ritual and Power in Postcolonial Africa, Chicago: University of Chicago.

Cornevin, R. (1959): Histoire du Togo, Paris: Académie des Sciences d'outremer.

Descola, P. (2005): Par-delà nature et culture, Paris: Gallimard.

Diouf, M./ Mamdani, M. (eds.) (1994): Academic Freedom in Africa. Dakar, Senegal: CODESRIA.

Dübgen, F./ Skupien, S. (eds.) (2015): Afrikanische politische Philosophie, Berlin: Suhrkamp.

Espinoza Garrido, F. (ed.) (2019): Locating African European Studies. Interventions, Intersections, Conversations, London: Routledge. https://doi.org/10.4324/9780429491092-1.

Essed, P. (ed.) (2019): Relating Worlds of Racism. Dehumanisation, Belonging, and the Normativity of European Whiteness, Cham: Palgrave McMillan. https://doi.org/10.1007/978-3-319-78990-3.

Gayibor, N. L. (ed.) (1997): Histoire des Togolais de 1884 à 1960. Deux Volumes, Lomé: Presses de l'Université de Lomé.

Geschiere, P. (1995): Sorcellerie et Politique en Afrique – La viande des autres, Paris: Karthala.

Glasman, J. (2008): "Review on Gayibor, Nicoué L. (ed.): Histoire des Togolais de 1884 à 1960 (Vol. 2). Lomé 2005." In: *H-Soz-Kult*, September 3, 2008 [online] (https://www.h-net.org/reviews/showrev.php?id=22788).

Hahn, H. P. (2001): "Fritz Graebner, Kulturkreise und Kulturschichten in Ozeanien, Bernhard Ankermann, Kulturkreise und Kulturschichten in Afrika." In: K.-H. Kohl/C. F. Feest (eds.), Hauptwerke der Ethnologie, Stuttgart: Kröner, pp. 137–147.

Holbraad, M./Pedersen, M. A. (2017): The Ontological Turn. An Anthropological Exposition, Cambridge: Cambridge University. https://doi.org/10.1017/9781316218907.

Horner, B./ NeCamp, S./Donahue, C. (2011): "Toward a Multilingual Composition Scholarship: From English Only to a Translingual Norm." In: College Composition and Communication 63/2, pp. 269–300. https://doi.org/10.58680/ccc201118392.

Hountondji, P. J. (ed.) (1997): Endogenous Knowledge: Research Trails, Dakar, Senegal: CODESRIA.

Kessi, S./Marks, Z./Ramugondo, E. (2021): "Decolonizing Knowledge within and beyond the Classroom." In: Critical African Studies 13/1, pp. 1–9. https://doi.org/10.1080/21681392.2021.1920749.

Kössler, R. (2015): "Solidarity with Liberation in Namibia. an Analytical Eyewitness Account from a West German Perspective." In: J. Sylvester (ed.), Re-Viewing Resistance in Namibian History, University of Namibia Press, pp. 252–265. https://doi.org/10.2307/j.ctvh8qxrv.22.

Kössler, R./Melber, H. (2002): Globale Solidarität? Eine Streitschrift, Frankfurt a.M.: Brandes & Apsel.

Leclerc, G. (1972): Anthropologie et colonialisme. Essai sur l'histoire de l'africanisme, Paris: Fayard.

Maldonado-Torres, N. (2016): Outline of Ten Theses on Coloniality and Decoloniality. Frantz Fanon Foundation (online) https://fondation-frantzfanon.com/outline-of-ten-theses-on-coloniality-and-decoloniality/ (Accessed 17 May 2024).

Mamdani, M. (2007): Scholars in the Marketplace: The Dilemmas of Neo-Liberal Reform at Makerere University, 1989–2005, Dakar, Senegal: CODESRIA.

Marboeuf, O./Ben-Yakoub, J. (2019): "Variations décoloniales. Conversation entre Olivier Marboeuf et Joachim Ben Yakoub (Mai 2019)." In: Toujours Debout, (Mai), pp. 1–16.

Misgav, H.-P. (2010): Décoloniser l'histoire occidentale. Les naissances politiques de l'anthropologie historique. In: L'Atelier du Centre de recherches historiques 6 [online] https://doi.org/10.4000/acrh.1914.

Mkandawire, T. (ed.) (2005): African Intellectuals: Rethinking Politics, Language, Gender and Development, Dakar, Senegal: CODESRIA. https://doi.org/10.5040/9781350218147

Ndlovu-Gatsheni, S. J. (2013): Coloniality of Power in Postcolonial Africa, Dakar: CODESRIA.

Ndlovu-Gatsheni, S. J. (2015a): "Decoloniality as the Future of Africa." In: History Compass 13/10, pp. 485–496. https://doi.org/10.1111/hic3.12264.

Ndlovu-Gatsheni, S. J. (2015b): "Genealogies of Coloniality and Implications for Africa's Development." In: Africa Development, 40/3, pp. 13–49.

Ndlovu-Gatsheni, S. J. (2018): Epistemic Freedom in Africa: Deprovincialisation and Decolonisation, London, UK: Routledge. https://doi.org/10.4324/9780429492204.

Ngũgĩ wa Thiong'o (1986): Decolonizing the Mind: The Politics of Language in African Literature, Nairobi: Heinemann.

Nkrumah, K. (1964): Consciencism. Philosophy and Ideology for De-Colonialization and Development with Particular Reference to the African Revolution, New York: Monthly Review.

Nyamnjoh, F. B. (2012): "Potted Plants in Greenhouses: A Critical Reflection on the Resilience of Colonial Education in Africa." In: Journal of Asian and African Studies 47/2, pp. 129–154. https://doi.org/10.1177/0021909611417240.

Nyamnjoh, F. B. (2015): "Introduction: Academic Freedom in African Universities." In: PAX ACADEMICA: African Journal of Academic Freedom/Revue Africaine de Liberté Académique 1–2/2015, pp. 7–14.

Nyamnjoh, F. B. (2017): "Incompleteness and Conviviality: A Reflection on International Research Collaboration from an African Perspective." In: Y. D. Gebre/I. Ohta/M. Matsuda (eds.), African Virtues in the Pursuit of Conviviality: Exploring Local Solutions in Light of Global Prescriptions, Bamenda, Cameroon: Langaa, pp. 339–378. https://doi.org/10.2307/j.ctvh9vwxv.14

Nyamnjoh, F. B. (2019): "Decolonizing the University in Africa." In: N. Cheeseman (ed.), The Oxford Encyclopedia of African Politics, New York: Oxford University. https://doi.org/10.1093/acrefore/9780190228637.013.717.

Probst, P. (2005): "Betwixt and Between an Anthropologist's Perspective on the History of African Studies in Germany." In: Afrika Spectrum 40/2005/3, pp. 403–427.

Quijano, A. (2000): "Coloniality of Power and Eurocentrism in Latin America." In: International Sociology 15/2, pp. 215–232. https://doi.org/10.1177/0268580900015002005.

Straube, H. (1971): "Die traditionelle Landwirtschaft Afrikas in historischer Sicht." In: Internationales Afrikaforum 7/8, pp. 449–454.

Uimonen, P. (2020): "Decolonising Cosmopolitanism: An Anthropological Reading of Immanuel Kant and Kwame Nkrumah on the World as One." In: Critique of Anthropology 40/1, pp. 81–101. https://doi.org/10.1177/0308275X19840412.

VAD (ed.) (1970): Interdisziplinäre Afrikanistik, Hamburg: Buske.

Part 2:
Decolonising the Academy – Practical Examples

"Because Rhodes Fell"

Historical development, institutional contexts, and the challenges of decolonisation of South African anthropology

Heike Becker

Figure 1: Rhodes statue removal, 2015. (Photo: Heike Becker)

Introduction

This chapter provides a critical discussion of the trajectories, institutional contexts, and challenges of decolonisation of South African anthropology. I present an in-depth exploration of the country's historical context and its implications for the discipline. The chapter's first part presents a concise history of anthropology in South Africa during the colonial, apartheid and early post-apartheid eras.[1] The second part reviews the 21st century challenges of decolonisation, anthropological research and teaching at different universities in the country.[2] The current changes, challenges and opportunities fundamentally resonate with the massive student protests which shook South Africa in 2015 and 2016. The paper's title references the key events that started the protests, namely the student activism to have a statue of Cecil John Rhodes, the British colonialist and mining magnate, removed from the campus of the University of Cape Town (UCT). I argue that these student-led movements, known as #Rhodes Must Fall, or #RMF, inspired a momentous surge to decolonise academic institutions and the curriculum in South Africa, and furthermore that they resonated also with students elsewhere, including in universities and public spaces in the Global North, epitomised in the Rhodes Must Fall campaigns at Oxford University, and also noticeable in the United States, for example, at Harvard (see Ahmed 2019).

A very brief history of anthropology in South Africa

As an exception to the situation on much of the African continent, in South Africa anthropology has been well-established since 1921. Today the discipline is taught at most universities in the country from undergraduate through to doctoral levels.

Anthropology in South Africa has solid institutional foundations, though departments of anthropology are generally small; the two largest anthropology departments, at UCT and the University of the Witwatersrand (known as "Wits"), have eight permanent academic staff members each, followed by a few other universities, including my own institution, the University of the Western Cape (UWC), with between five and seven lecturers; several universities only have one or two teachers of anthropology. South African anthropology has a longstanding history of professional associations, conferences and academic journals.

South African anthropologists have been influential in the international development of the discipline, particularly in the Anglophone Global North. In

1 A more detailed history of South African anthropology is presented in Spiegel and Becker (2018).

2 Some of the material presented in this chapter is concurrently also being published in a forthcoming article in *Sociologus-Journal for Social Anthropology*.

the mid-20th century South African-born, and a few South African-based anthropologists, including Meyer Fortes, Isaac Schapera, Max Gluckman, Hilda Kuper and Monica Wilson, were critical for the development of British social anthropology. More recently the work of South African-born anthropologists, prominently Jean and John Comaroff, and, increasingly, South African-based anthropological research has been noted in the North American academy.

In the 21st century, anthropology in South Africa faces challenges that originate in the country's past, as well as in the political and institutional changes of the post-apartheid period. South African anthropology has had to face the critique of anthropology as a handmaiden of colonialism both from within and beyond the discipline. Notable criticism was raised by Black[3] South African anthropologists in the early 1970s (Mafeje 1971; Magubane 1971, 1973). In a widely discussed intervention as late as the mid-1990s, Archie Mafeje, one of South Africa's first Black trained anthropologists, suggested that the discipline should "commit suicide" (Mafeje 1996).

Concerns about South African anthropology have continued to be raised during the post-apartheid era. Over the past two decades the face of South African anthropology has changed quite dramatically. What was until fairly recently a small disciplinary community and demographically almost exclusively white, has since the 1990s significantly expanded and diversified. Today anthropology programmes train post-graduate students from diverse backgrounds, including, overwhelmingly, Black research students at some historically black universities, and increasing numbers of Black South African anthropologists, as well as some scholars who hail from elsewhere on the continent, and who have been appointed to academic teaching posts. Nonetheless, the racial demographics remain uneven across universities. Inequalities are not only based on racial difference, even though this can never be neglected in the South African context. The leading, historically White universities, such as UCT, Wits, or the historical Afrikaans-elite institution in Stellenbosch, have

3 A note on the use of the terminology in reference to racial categories: In contemporary South Africa, "African" and "black" are commonly used interchangeably. Black, with a capital B, in contrast refers to an inclusive, political usage to include all people, who have been historically oppressed because of racialisation, i.e., "black", "coloureds" (people of mixed-race descent), and "Indian" (descendants of 19th century immigrants and indentured labourers from the Indian subcontinent). I refer to "whites" generally in the lowercase, except in specific historical contexts where the term was previously used by the apartheid regime to denominate "Whites-only" institutions.

become attractive for the small, but growing numbers of Black elite and middle-class families who can afford the considerably higher tuition fees at these universities, whereas universities such as UWC still draw in mostly students from Black working class and rural backgrounds.

Material inequalities between historically unevenly resourced institutions persist and these impact, for instance, on the funding opportunities for postgraduate students, and institutional research funding. Much needed funding is more readily available at the better-resourced historically White universities than at the under resourced historically Black universities. Also, while undergraduate class sizes are generally high across South African universities, classes are even larger at those universities that attract mostly students from poor, Black backgrounds. In addition, such students are usually less well-prepared for academic studies due to the immense inequalities of the country's education system.

Institutional contexts and demographics remain significant. Furthermore, any meaningful discussion of anthropology in South Africa needs to take into consideration that from its inception as a formally recognised academic discipline in the early 20th century, anthropology has been acutely implicated in the country's political-economic history (Spiegel and Becker 2018). The ideal of a purely "scientific" comparative study of human society and culture that some anthropologists may have striven toward has thus never been realisable. The context in which South African anthropology has developed has, from the start, been one of quite radical and thoroughly political circumstances. South African anthropology has been shaped by, and in turn has sometimes influenced dynamic socio-economic and cultural processes. These include rapid industrialisation, the effects of a strong British imperial colonising presence and resistance to it by indigenous people, and the presence of a long-time resident European settler population. The trajectories of South African anthropology have been subject to racialised capitalism, and a historically constituted web of uneven and unequal ties connecting South Africa as a sub-imperial metropole with the southern African region as a whole. These have included transnational (in addition to national) labour migrancy, economic and political dependency, violent interventions in neighbouring countries, and – in the case of Namibia – formal colonisation.

Beginnings: the "native question"

The first anthropology programme in South Africa was formally established in 1921. It was founded, under A.R. Radcliffe-Brown, at UCT, the country's oldest university, which was established in the British tradition, and attended by an overwhelmingly white student population until the 1990s when the institution's demographics began to change. UCT Anthropology started in the School of African Life and Languages, following calls from the European settler population for the support and encouragement of "scientific" study to address what was considered "the native question." (Spiegel and Becker 2018) The "native question" had long preoccupied the minds of colonial governors and missionaries, such as Henri-Alexandre Junod (1863–1934), a Swiss Christian missionary-turned-ethnographer. While Junod attempted salvage ethnography, government administrators, were concerned with "the native question" from a policy perspective.

Chairs of Bantu Life and Languages were soon also established at the Afrikaans-medium Stellenbosch University, the University of the Witwatersrand, and the University of Pretoria. The political economy context was decisive, at Wits the anthropology programme tellingly being supported by two mine-labour recruiting agencies within the Chamber of Mines. Alongside social anthropology and Bantu linguistics, Wits also offered classes in native law and administration (Spiegel and Becker 2018).

We thus need to keep in mind that those seeking to establish social-cultural anthropology in South Africa saw it as a tool for understanding and managing the imposition of modernity and industrial labour on indigenous South Africans. The first defining characteristic is that, despite occasional protestations that their discipline was a distanced "science" of social and cultural diversity conducted for its own sake, South African anthropology was of a distinctively applied nature from the very beginning. This has remained true throughout the century of its existence and has defined the professional framework of most local anthropologists. Regardless of their varying political and epistemological orientation, South African anthropologists have been driven in their work by the political-economic context in which they found themselves (Spiegel and Becker 2018).

Volkekunde: paradigmatic apartheid ideology in South African anthropology

Secondly, South African anthropology has been marked by division, to the extent that it has sometimes been asked whether we can indeed speak about "South African anthropology" at all, or should rather refer to a plurality of anthropologies (Spiegel and Becker 2015). For much of the 20th century anthropology in South Africa was characterised by the opposition between social anthropology, which was closely connected to its British counterpart, and *volkekunde*, the nationalist Afrikaner version of the discipline. These different strands, and particularly the entanglement of *volkekunde* and apartheid were subject to a substantial literature on anthropology and apartheid in the 1980s and early 1990s (e.g., Sharp 1981; Gordon and Spiegel 1993). Although this emphasis was to an extent owed to the political and social context of the late apartheid era, a concise discussion of the differences – and the commonalities – of these two branches is required if we want to understand the historical development of anthropology in South Africa.

If we first take a look at the *volkekunde* branch of South African anthropology, the approach's nationalist and racist implications are of imminent urgency, as are questions of pedagogy and habitus in Bourdieu's understanding. The historical auto-ethnography of C.S. (Kees) van der Waal provides a fitting starting-point for this exploration. Van der Waal was trained in the *volkekunde* paradigm at the then staunchly Afrikaner-nationalist University of Pretoria but had already turned his back on the epistemological community by the late 1970s, during the high apartheid era. He retired in 2015 from his professorial position at Stellenbosch University. In his valedictory lecture, published in *Anthropology Southern Africa,* he revisited his personal, political and intellectual trajectory from his 1950s Pretoria childhood. He depicted the bleak atmosphere at the Afrikaans university of Pretoria, writing that, "becoming a student at UP in the 1960s was like entering an ideological trap where Afrikaner ideology crept into every corner of the social sciences and humanities" (Van der Waal 2015: 220). He pointed out that the anthropology student body was predominantly male, and at the Honours (junior postgraduate) level dominated by government officials in charge of the "natives". University structures and pedagogy were emphatically authoritarian. Critical thinking, or even just wider reading were discouraged (van der Waal 2015: 221–22). This was Pretoria University in 1968, the year when students around the globe, including at some South African universities, rose up in anti-authoritarian revolt (Becker 2018).

Pretoria's *volkekunde* department was dominated in thoroughly patriarchal manner by the Coertzes. Father Pieter J. ("Piet"), a student of the *volkekunde* founder, Werner Eiselen, at Stellenbosch University in the 1930s, had taught in Pretoria from the 1950s. He was later joined by his son and successor in the chair, Roelof. Van der Waal (2015: 223) remembers the personally and politically stern presence of these men, accompanied by their "deep sense of worry about 'race relations' and the lack of political will of the government and electorate to move towards total separation". The research they did focused on the documentation of indigenous law, which they saw as their contribution to the strengthening of traditional authorities in the Bantustans. They were also active outside the university under the banner of *volksdiens* ["service to the (Afrikaner) people"; HB]. Formality in language, formal meetings and a strong sense of control were all part of the way the Department was organised (ibid.).

Control and formality also dominated the ways in which research was conducted. Deep immersion during fieldwork was discouraged. Instead, "their mode of fieldwork often entailed formal interviews in tribal offices where designated old men would present the indigenous legal system, based on a research schedule that had been developed for the replication of several projects" (Van der Waal 2015: 223).

Conceptually, the Afrikaner-nationalist anthropologists emphasised the significance of "culture", understood as cultural difference. In the student textbook, developed under the leadership of Pieter J. Coertze (1959), they presented their discipline as a study of singular, unified, and historically persistent groups of people and their respective and distinctive cultures. *Volkekunde* emphasised that humans were members of culturally separate peoples, and that each of these lived according to their culture in an integrated *ethnos*. Each *ethnos* was demarcated with clear boundaries, and members of each new generation were enculturated into it. "Culture contact" with others in the wider South African social context was regarded as an immense danger to the naturalised and deep cultural differences that were presumed to exist between people, classified differently in terms of race, language and culture.

With the end of the apartheid dispensation, *volkekunde* quickly disappeared from the scene. Its decline had started earlier. In the case of one white-Afrikaans university's anthropology department, at the then Rand Afrikaans University (today known as the University of Johannesburg), the entire lecturing staff explicitly turned their backs in the early 1980s already on the paradigm in which they had all been trained. At Stellenbosch, where *volkekunde* had been devised under Eiselen in the 1930s, the department was closed down in the mid-1990s. The post-apartheid trajectory of Stellenbosch anthropology is interesting. A few years after the closure of the university's *volkekunde* department, social anthropology was introduced within the Department of Sociology. A protagonist of the decidedly Marxist approach that had become influential in South African social anthropology in the 1980s (see below) was appointed to the chair. Today one would be hard pressed to find any anthropologist in South Africa who would self-describe as a *volkekundige*.

Social Anthropology: from the "native question" to exposé

They would not have liked to hear of it, but the social anthropology taught in South African English-medium universities for many years bore distinctive similarities to *volkekunde*. The structural-functionalist approach that was dominant since Radcliffe-Brown's appointment in 1921 emphasised static models of African lives determined through membership of timeless, bounded cultural units. Social anthropology, too, regarded "culture contact" between Europeans and Africans as disrupting this "natural state", and resulting in the deplorable "detribalised native". Implicitly, at least, this approach still persisted – and was heavily critiqued – in the late 1950s *Xhosa in Town* trilogy of Philip Mayer and his colleagues. (Mayer and Mayer 1971; Magubane 1973)

However, in contrast to the monolithic *volkekunde* branch, this conservative approach was repeatedly challenged by younger social anthropologists who subscribed to more critical or even radical interpretations of the discipline. By the 1940s some social anthropologists had stopped worrying about "detribalisation". South African anthropologists associated with the Rhodes-Livingstone-Institute (RLI) in Northern Rhodesia (today's Zambia) focused on the broader, political and economic structures of changing African lives (see, e.g., Gordon 2018). As early as 1939, the institute's first director, Godfrey Wilson, who was British and married to the (later) influential South African anthropologist Monica Hunter (Wilson), stated programmatically: "It is with an Africa no longer primitive, held fast in the economic system of a world society, partially Christian, ruled by Europeans with which the student must begin and end" (Wilson 1939: 5). Two decades later, the then RLI director, Max Gluckman, who was born in 1911 in Johannesburg to Eastern European Jewish immigrants and trained in anthropology at Wits in the early 1930s, pronounced famously that "an African townsman is a townsman, an African miner is a miner" (Gluckman 1960: 57).

Gluckman never taught at a South African university. However, he remained influential because of his radicalisation of Radcliffe-Brown's earlier assertion, made during his inaugural lecture at UCT in 1922, that South Africa had to be studied as a single system, with blacks and whites as component parts. No "culture" (or "society" in the structural-functionalist lingo) could be studied as a separate entity. Gluckman, among others, went further with their emphasis that African "societies" could not be understood outside the history of colonial conquest and a racist society (Gluckman 1940; Hunter 1936).

During the late apartheid years, a political economy approach became even more influential, gathering momentum from the late 1970s onwards. Historians, social anthropologists and sociologists promoted Marxist scholarship first in the Wits History Workshop, a radical interdisciplinary space. The strongest foothold of Marxist approaches in social anthropology however was at UCT where a generation of lecturers and students turned their interests to the devastating consequences of mass relocation in the country (Spiegel and Becker 2018).

This school of South African social anthropology became known as "exposé anthropology" as it was "designed to demonstrate many of the worst on-the-ground consequences of the apartheid system", as one of its chief protagonists later explained (Spiegel 2005: 133). Poverty and processes of social stratification emerged as the central concerns, along with influx control, Bantustans, and the politics of ethnic nationalism.

In their rejection of apartheid politics and the corresponding claim to ontological differences between "cultures" in the *volkekunde* brand of cultural anthropology and – implicitly – their structural-functionalist teachers, South African anthropologists of the late apartheid period, deviated from old school anthropology. They shied away from an examination of anything deemed "culture" (Gordon and Spiegel 1993: 87). As some have pointed out since, South African social anthropologists of that era showed little concern with local meanings of the transformations wrought by apartheid, including the expressions of popular culture which had emerged in response to it (James 1997: 116; Bank 2011: 262; Becker 2012 passim; Van Wyk 2012 passim).

Nonetheless, it should be noted that the Marxist-influenced anthropologists unpacked in one remarkable project the genealogy of the segregationist and apartheid terminology, notably "culture", "community", "race", or "ethnic group" and "nation" (Boonzaier and Sharp 1988). The resulting publication (ibid.), *South African Keywords*, was probably more widely read and prescribed on South African social anthropology and other social science courses than any other publication in the late 1980s and early 1990s. Although, with one exception,[4] the editors and contributors were all members of the white section of South Africa's population, "*Keywords*" was also widely read in activist anti-apartheid circles.

Who were South Africa's anthropologists of the 20th century?

With a few exceptions, virtually all practicing anthropologists and most students in both social anthropology and *volkekunde*, belonged to the white settler population until well into the final apartheid years. However, certain significant differences of gender and ethnicity need to be noted between the different disciplinary orientations.

As Andrew Bank's studies of Werner Eiselen and his students have shown, almost all *volkekundiges* were male, and generally of upper middle-class Afrikaner background (Bank 2015). An overarching masculine environment is also remembered by Kees van der Waal (2015) from the late 1960s through to the 1980s.

In the social anthropology tradition, on the other hand, there was a strong presence of women, whose significant contributions have recently been highlighted by Bank (2016). Winifred Hoernle, who, as Bank argues, should be considered the "mother" of South African anthropology and the generation of her students, including, notably, Monica (Hunter) Wilson, Eileen Krige, Ellen Hellmann and Hilda Kuper, contributed remarkable ethnographies. Although all of them were white, South Africa's women anthropologists of the 20th century varied in their social and ethnic backgrounds. Monica Hunter (later Wilson) grew up as the daughter of Christian missionaries in the Eastern Cape where she later conducted her first, famous fieldwork for the classic, *Reaction to Conquest* (Hunter 1936). Eileen Krige married Smuts' nephew Jack Krige and was thus affiliated to the liberal Afrikaner tradition. Ellen Hellmann, who conducted the first African in-depth urban ethnography in the mid-1930s (Hellmann 1948) was of European Jewish immigrant background. So were Hilda Kuper and a number of influential male anthropologists, including Isaac Schapera, Meyer Fortes, and Max Gluckman, all born in the early 20th century to recent Jewish immigrants to South Africa.

4 Mamphela Ramphele, a leading former Black Consciousness activist, who was then a member of the Social Anthropology Department at UCT.

Although the discipline was overwhelmingly white, a few Black intellectuals trained in social anthropology. Most prominent among them was Z.K. (Zachariah Keodirelang) Matthews, who was born in 1901 near Kimberley as the son of a mineworker. He attended a mission high school in the Eastern Cape, where he then also studied at the South African Native College (the predecessor of Fort Hare University) from 1918. In 1924 Matthews became the first black ("African") South African to earn a Bachelor's degree from a South African institution of higher learning. In 1933 he earned a LLB degree, then, at Yale in 1934 he completed his master's thesis, *Bantu Law and Western Civilisation in South Africa: A Study in the Clash of Cultures*, and finally studied for a year in London under Malinowski who regarded him as one of his most talented students. In 1935 Matthews returned to South Africa and a year later was appointed as a lecturer in Social Anthropology and Native Law and Administration at the University of Fort Hare. In 1944 he was promoted to professor and became Head of Fort Hare's Department of African Studies, where he was joined by Monica Wilson. They became close colleagues and friends, a relationship that continued throughout their lives.[5]

At Fort Hare, Matthews and Wilson trained a small cohort of Black South African anthropologists during the 1940s, including, among others, Livingstone Mqotsi and Godfrey Pitje. The political situation of the time, however, precluded any of them from pursuing a professional career in anthropology. Mqotsi, for instance, went through a string of jobs but never found appropriate employment as an anthropologist although he was finally awarded an MA by Wits in 1957 for his dissertation, *A study of ukuthwasa: (being a syndrome recognized by the Xhosa as a qualification for being initiated as a doctor)* (1957). He went into exile and ended up teaching history at a Comprehensive (high school) in the United Kingdom. During that time, he also published an ethnographic-political novel, *The House of Bondage* (1989) about life and struggles in the Bantustans.

5 Matthews was a leading activist in the ANC and in the Defiance and Freedom Charter campaigns of the 1950s. He was among the accused in the Treason trial. After being acquitted in late 1958 he returned to Fort Hare, but soon afterwards resigned his post in protest against the passing of legislation that reduced the university to an ethnic college for Xhosa. He subsequently left South Africa in 1962 and died in 1968 while serving as ambassador of Botswana to the United States. Monica Wilson contributed substantially to his autobiography, which was published posthumously: *Matthews, with a Memoir by Monica Wilson* (1981).

At the time when the first wave of decolonisation swept across most of the African continent, the opposite happened at the southern tip of the continent with the implementation of the rigid settler colonial apartheid dispensation. The hardened political stance also hit the South African academy and closed off the already minute spaces of academic opportunity for Black students and scholars. From 1959 onward, when the infamously misnamed Extension of University Education Act was passed, South African students had been admitted to universities strictly along racial and ethnic lines.

This had a significant impact on anthropology as well. Promising Black scholars, such as Bernard Magubane and Archie Mafeje were forced into exile. The experience of Archie Mafeje, Monica Wilson's star student and co-author in the early 1960s, is particularly poignant (Wilson and Mafeje 1963). In 1968 Mafeje, by then in the process of completing his PhD at Cambridge University, was appointed to a senior lecturer position in Social Anthropology at UCT. The university offered him the job, but then, following government pressure rescinded the offer. This led to one of the best-known student protests of South Africa's "1968 moment", a "sit-in" occupation of the university's administration building, which was however ultimately unsuccessful. A white anthropologist was appointed in Mafeje's place. South Africa's oldest university had caved in to the demands of the apartheid policy regarding university education. The 1968 Mafeje affair must be understood as representative of the enforcement of apartheid policies in the academy.

At UCT, which had been declared a White institution under the infamous 1959 Act, Black students were admitted only under exceptional circumstances and any "non-white" applicant aspiring to study at UCT had to apply for a special permit from the government. Although this law did not pertain to academic staff members, Mafeje's appointment was prevented (Becker 2018: 35–37). Mafeje took up a series of appointments in various African and European universities and became known for his critique of anthropology. In the late 1990s he participated in vibrant discussions held among African anthropologists about the future of the discipline on the continent.

Hence, until many years later, virtually all anthropologists in South Africa were white. Before the 1990s, only a handful of Black scholars and students entered the sought-after programmes at universities such as UCT and Wits. There were few exceptions, among them, Mamphela Ramphele, already a practicing physician and a prominent activist of the Black Consciousness movement, who embarked on a research project and doctoral degree with the UCT Social Anthropology department after she had been awarded a grant from the Carnegie Foundation in the mid-1980s. Even where they were able to contribute, Black anthropologists, and even white women, though not quite to the same extent, have until recently been largely absent from disciplinary histories (Bank and Bank 2013; Bank 2016).

In post-apartheid South Africa, however, the discipline has enjoyed robust student intake, and there has been a steep rise in numbers of postgraduate students. The disciplinary community, previously concentrated in a handful of universities and demographically almost exclusively white, has significantly expanded and diversified. Today many anthropology programmes train postgraduate students of diverse backgrounds, including large numbers of Black research students at some historically Black universities. Nonetheless, the racial demographics of the anthropology student population remain uneven across universities (Spiegel and Becker 2018).

The almost exponential growth in postgraduate student numbers since the late 1990s has, in part, resulted from institutional pressures in the face of international university ratings systems and the increasing demand for senior qualifications in order to enter the labour market.[6] It has also been driven by the increasing number of international students, predominantly from other parts of Africa and the Global South, registering at South African universities. The international African students have joined a steady though still limited increase in the number of Black South African postgraduate students entering anthropology programmes.

Professional organisations of Anthropology

A long-established trajectory of disciplinary conferences and associations has contributed to the perception of one, or multiple, distinctly South African anthropology / anthropologies. From 1967 the South African anthropology conferences which were held annually were informally supervised by a committee. These conferences took always place at the distance-learning and, at that point, bilingual (English and Afrikaans) University of South Africa (UNISA). Despite the tensions between the social anthropology and *volkekunde* traditions, conference participants came from both disciplinary branches. However, in 1977, after a few black anthropologists began attending the conference, the *volkekunde*

6 Countrywide, total university enrolment increased from just under 500,000 in 1994 to twice that ten years later (Spiegel and Becker 2018).

camp withdrew and established its own association, known by its Afrikaans-language name as the *Vereniging van Afrikaanse Volkekundiges* (VAV; Association of Afrikaans Ethnologists), later renamed South African Society for Cultural Anthropology (SASCA).

The annual anthropology conference continued informally for another ten years until, in 1987, another association entered the scene, known as the Association for Anthropology in Southern Africa (AASA). AASA was open to all anthropologists, however it required aspirant members to subscribe to a clause in its constitution that expressed an explicit rejection of apartheid. In effect, AASA thus brought together the liberal and Marxist adherents of social anthropology at the English-medium universities, a few renegades from the Afrikaans-medium institutions, and those teaching at the universities for Black students, which had been created by the apartheid state in the 1960s, including a very small number of Black scholars. Anthropology at the ethnic universities (following the 1959 misnamed Extension of University Education Act) had generally started with adherence to the *volkekunde* paradigm but, in the course of the 1980s, more critical perspectives had developed at some of these institutions. The University of the Western Cape, for instance, originally designated as the ethnic university for coloured students had turned to radical scholarship and activism in the 1980s; by the time apartheid came to an end, UWC anthropology lecturers included, among others, some who subscribed explicitly to Marxist and feminist approaches.

In the mid-1990s, some, though by no means all, members of the Afrikaans anthropology association started attending the annual AASA conference. A decade into the democratic post-apartheid dispensation, those linked with each of the two branches of the discipline came together to form a new unified association, which became known as Anthropology Southern Africa (ASnA). ASnA held its first annual conference in 2001. Since then, the annual ASnA conferences have been hosted at various university campuses around South Africa. In the later 2010s conferences also took place elsewhere in the wider southern African region, including in Malawi (2017) and Botswana (2018). The 2020 conference was to be held at the University of Namibia in Windhoek, although it was cancelled due to the Covid pandemic.[7]

In contrast to the earlier associations that were confined to anthropologists working in South Africa, ASnA aspires to organise anthropologists through-

7 Previously an AASA conference was held in Zimbabwe in the late 1990s and another one in 2000 in Namibia.

out the entire southern African region, including Angola, Botswana, Lesotho, Madagascar, Malawi, Mozambique, Namibia, South Africa, Swaziland, Zambia and Zimbabwe. The association has also sought international affiliations. It was a founding member of the World Council of Anthropology Associations (WCCA) formed in 2004 in Brazil. ASnA retains its links with the WCCA, as well as the International Union of Anthropological and Ethnological Sciences, the Pan African Anthropological Association (PAAA) and is in the process of connecting more closely with emerging anthropology networks in different parts of the continent.

A journal for Southern African anthropology

Historically, South African social anthropologists published mostly in British journals, and locally in interdisciplinary African Studies journals, particularly the Wits-based journal *African Studies* (originally known as *Bantu Studies*). When the two South African anthropology associations merged in 2001, the VAV brought their journal, then published bi-lingually as the *Suid-Afrikaanse Tydskrif vir Etnologie / South African Journal for Ethnology*, into the marriage. Renamed, *Anthropology Southern Africa*, the journal slowly opened up to more inclusive positions. From 2011 the publication's transformation accelerated, and the journal was relaunched with a new vision and new editorial board in 2014. The "new" ASnA is envisioned as a transnational journal that is firmly based within, and speaking from southern Africa, while also reaching out to international scholars, who are engaged in southern African scholarship. The new concept was intensively discussed within the association's executive, an interim editorial collective and at the annual ASnA conferences, and eventually put to the vote of the membership. The membership overwhelmingly supported the proposed internationalisation and the journal has since been published through Taylor & Francis.

A significant challenge was to move from a previously solely South African-focused publication to a journal whose commitment is to a *Southern* African reach in terms of published content as well as readers, authors and editors. Over the past few years much has changed, most visibly embodied in the journal's diverse editorship; recently editors have been based in the wider Southern African region, and even in Europe (Switzerland). During 2021 and 2022, the journal's editorial team was headed for the first time by an editor-in-chief from the southern African region beyond South Africa, namely Romie Nghitivelekwa from the University of Namibia (UNAM).

The challenges of regionalisation were rooted, partially, in the disciplinary challenges anthropology has been facing in much of postcolonial Africa. Anthropology is currently not formally taught in several Southern African countries, including Namibia, Botswana, Malawi, and Swaziland (although there are currently moves underway towards establishing anthropology programmes in some of these countries).[8] On an intellectual level, the challenges have entailed a move away from "South African exceptionalism" towards an ethnographic theoretical analysis of southern Africa as an integrated region that has been constituted through a web of uneven and unequal historical and contemporary ties.

"Because Rhodes fell": Decolonising the anthropology curriculum

In 2017 Mahmood Mamdani, decolonial scholar at Makerere University and Professor of Anthropology at Columbia University, accepted an invitation by UCT to deliver the university's annual T.B. Dawie Memorial Lecture on academic freedom. This was an extraordinary occasion. Mamdani had previously headed UCT's Centre for African Studies between 1996 and 1999, and had left, in a state of frustrated protestation, an institution that he considered hopelessly untransformed, vowing never to return. When he came back two decades later, he responded to the obvious questions about why he had changed his mind, saying that he could now return to UCT "because Rhodes fell".

Mamdani's response signifies a view held by many that there is now a space in South African universities for robust conversations about issues that were previously not raised in the post-apartheid South African academy. Questions about the politics of knowledge and curriculum reform were forcefully put on the agenda by the massive student movements that rocked the country in 2015 and 2016. These protests need to be considered as a significant catalyst for the South African efforts of decolonising institutions and curricula.

In March 2015, students at UCT had begun a forceful campaign, dubbed #RhodesMustFall (#RMF) to have the statue of the British colonialist and mining magnate Cecil John Rhodes removed – this monument had been sitting on

8 Following several years of planning, UNAM offers undergraduate, MPhil and PhD programmes in Anthropology from 2024.

the university grounds for the past eighty years. Activists who successfully disrupted "business as usual" on the UCT campus occupied the university's main administrative building, initiated a debate about racism and voiced their demands to decolonise education. The movement succeeded in gaining the support of the university's governing bodies; on 9th April 2015 the objectionable statue was removed under the thunderous applause of a large crowd who had gathered to watch this significant moment (Becker 2018).

The movement spread quickly to other universities, initially mostly the historically white English-medium institutions with deep roots in British colonialism. As has often been pointed out, these universities' corresponding institutional cultures regularly alienate Black students (e.g. Naidoo 2016). In the second half of the year, the protests extended further so as to include the historically Black universities, which are today mostly attended by Black working-class students. At the Black universities, the immediate financial issues, such as tuition fees, often provided the trigger for protests, while discussions about academic canons and the racial composition of the professoriate came up later. In the end, though, a broad radical approach took hold across the spectrum, which may be summarised with the words of Nelson Maldonado-Torres who, in Fanon's spirit, beautifully characterised the events as efforts "at rehumanizing the world" (2016: 10).

By October 2015, the entire country was in the throes of mass protest. Initially, the protesters were voicing their opposition to the government's plan to increase tuition fees at public universities, but then they started calling for "free education". Primarily, this meant the abolition of tuition fees, but ultimately it was a call for far more significant change. Students were demanding no less than the intellectual and political liberation of a post-Apartheid society that the young activists perceived as still being profoundly racist. Drawing on the ideas of anticolonial thinkers, such as Frantz Fanon, and the "Black Consciousness" ideology espoused by Steve Biko, the South African activist murdered in 1977, the students called for the "decolonisation" of South African society.

"Decolonisation" became the catch word of the movements. While the demographics of most institutions of higher education have changed dramatically since 1990 and South African universities today generally have a black majority among their student body (though not among their academic staff), their institutional cultures, symbolism, and curricula have changed only marginally. This became a crucial issue for the new South African student movements. Decolonising institutions, decolonising knowledge, decolonising the mind, became the tags of the new generation of activists.

At universities across South Africa, groups were formed to explore ways to realise decolonisation in practice. Students campaigned for an overhaul of the symbols that embodied universities' institutional culture, and also called for the removal of controversial monuments and for buildings to be renamed. The call to recruit more Black teaching staff grew louder. There were also demands to reform curricula, which the students felt often perpetuated racist and colonial forms of knowledge while ignoring African traditions of learning and philosophy.

Decolonising the South African academy

The above notes about the connections of activism, transformation and the academy allow us to understand the ways in which the "class of 2015" introduced "decolonisation" into the discourse of South African universities. While "decolonisation" has been part of the language of anti-colonial struggles for decades, in education it was not a common referent; earlier discourses had referred instead to "liberation pedagogy", or "transformation". These changes beg the question of why the term suddenly erupted into political and academic discourse. What does decolonisation actually mean in the context of postcolonial South African politics of knowledge? And what does it mean in anthropology specifically?

It makes good sense to start the discussion by recalling the lecture delivered in 2017 by Mamdani at UCT. Titled, *Decolonising the Post-Colonial University* Mamdani's lecture emphasised that the African university "began as a colonial project – a top-down modernist project whose ambition was the conquest of society. The university was in the front line of the colonial 'civilising mission.'" Its aim was to create "universal scholars" who stood for excellence regardless of context – "the vanguard of the 'civilising mission'", as Mamdani (2019: 17) states without hesitation.

Evidently Mamdani illustrated his discussion with recourse to debates on the project of the (post)colonial university that had been raging in the early 1970s between different groups of academics affiliated to Makerere University in Uganda and the University of Dar es Salam ("Dar") in Tanzania respectively. His reflections on the postcolonial African university had fulminated about disparate visions of higher education: "excellence" versus "relevance"; the universal scholar "fascinated by ideas" versus the committed public intellectual. (Mamdani 2019: 18)

Mamdani, who was a student activist at Dar at the time, promotes a dialectical approach that acknowledges that each position contained something of value: having stated that context "obviously" matters since knowledge production is never immune from power relations, he acceded, yet cautioned: "At the same time, ideas also matter. If they did not, why have a university at all? This is to say that politics is not all". (Mamdani 2019: 18)

The challenges of anthropology in the contemporary South African university are inevitably framed by the transformation that happened in the post-1994 academic environment, as much as that which did not come to pass. Researchers in higher education, for example, Le Grange (2019), have pointed out that until universities were pushed by the 2015–16 movements, critical engagement with knowledge production and pedagogy was shoved to the margins of the higher education sector.

During the first two decades of post-apartheid South Africa, the emphasis of "transformation" of the academy was on the "massification" of higher education. The increase of black ("African") students' enrolment was the central aim. Significantly this took place within the ascendancy of neoliberalism and technologies of performance in post-apartheid South Africa, and particularly in the academic sector (Le Grange 2019: 30). State subsidies to the higher education sector declined severely (Le Grange 2019: 34). Insufficient state funding for South African public universities caused the steep increase of student fees as well as the outsourcing of auxillary services and the appalling deterioration of labour conditions for cleaners, security staff and other low-paid campus workers (Becker 2019). Academic work was subjected to a hollow idea of "excellence" and mechanical measurements of performativity. A peculiar system of output-dependent research subsidies was prioritised along with something of particular concern for anthropologists, namely an obsession with rather mechanical "ethics" procedures that often inhibited rather than supported ethically sound research practices.

Little attention was paid to what happened in the classroom in terms of content and pedagogy. The denigration of local, and generally non-western epistemologies, referred to by some scholars such as the historian Sabelo Ndlovu-Gatsheni (2018) as "epistemicide" continued quite unabashed. Little was done to change the hierarchical nature of student-teacher relationships, or the commonly-held notion that academic teaching would primarily involve transfer of knowledge rather than opening up questioning minds. In short, some progress was made in terms of desegregation and access to higher learning for previously excluded sections of the population. However, this progress was offset by the presence of inherited vestiges as well as new manifestations of neocolonialism, and by the rise of the neoliberal university under conditions of austerity: both features typified the realities experienced in post-apartheid South African universities.

Toward decolonial anthropology?

With the fall of (the) Rhodes (statue) the parameters of South African anthropology shifted. Anthropology students and even some of their lecturers were among those at the forefront of the protest movements. Strident demands for decolonisation were no longer just the preserve of student campaigners. Mostly, but not exclusively, younger and Black anthropologists put relevant questions on the agenda of the journal, the annual conferences, departmental and institutional meetings. Last but not least, they raised questions in the classroom.

Most South African-based anthropologists nowadays agree on the discipline's colonial legacy, namely "that the positioning of Africa and Africans for the purpose of study and analysis was historically tied to the grand narrative and experimentation of European colonization" (Boswell 2017: 4). Ultimately, a historical and colonial anthropology juxtaposed the observers, analysts and knowledge producers (anthropologists) onto the observed and analysed (African) subjects. Typically, knowledge production "solidified discourses of unequal abilities and social distance" (Boswell 2017: 4).

The social oppositions between "knowers" and "subjects" continue to reverberate in contemporary South African anthropology. The question: "Why have the affluent white sections of the South African population found little anthropological attention?" was asked a decade ago by Francis Nyamnjoh, the Cameroon-born Professor of Social Anthropology at UCT (Nyamnjoh 2012). His intervention stirred a robust debate (see e.g., Hartnack 2013; Niehaus 2013). In this situation it comes as no surprise that controversies tend to roughly follow the lines between those who study "their own" and those who study "the other".

Some anthropologists from marginal backgrounds in global hegemonic contexts have promoted doing "anthropology at home" to resolve this conundrum. The Nigerian anthropologist Wale Adebanwi, for instance, characterised his ethnography of Yoruba elites as the experience of the "native anthropologist", who is "speaking with and for, as well as acting as part of, 'them'." (Adebanwi 2017: 164) Similarly, the late South African feminist anthropologist Elaine Salo engaged in extensive reflection on her role as the "'native anthropologist'" (Salo 2018). At the time of her fieldwork in the late 1990s this was a significant epistemological breakthrough in South African anthropology. Salo conducted her fieldwork on the Cape Flats in ways "in which the field was home". She did not have to negotiate access linguistically since she already spoke the local language, Kaaps (a version of Afrikaans spoken in Cape Town's townships). Salo also had prior knowledge of the area since she had worked there with anti-apartheid activists in the 1980s; she was regularly updated on local issues through a network of public intellectuals who lived and worked in the area, including her brother who served as the local Anglican parish priest.

In the significant historical context of the South African politics of difference Salo shared the racial categorisation of the people she worked with. Yet, her lived experience as the daughter of a well-to-do family, a graduate of the historically White UCT, and her connections with people whom the ordinary residents of the township considered "insider-outsiders" also set her apart. Salo's acute awareness of the situatedness of knowledge production, which she regarded as inevitably enmeshed with classed, racialised and gendered experiences and locations, resulted in ethnographic work that is remarkable for compassion and sensitive relationships with the people with whom she worked in her research on the production of gendered and racialised personhood (Salo 2018).

This nuance distinguishes Salo's argument from those who emphasise a notion of Black anthropologists' studies being fundamentally different in outlook from those of white researchers simply on the basis of race. There is no doubt as to the legitimacy, even necessity, of struggling with and "writing against" the powerfully othering legacy of colonialism and apartheid. Yet in some presentations, the African-nationalist line of argument leans towards nativism that manifests in a search for an essential African-ness on which to build a foundation of black African dignity. The African-nationalist school of thought has thus far found little support among teaching academics. However, it has been growing among some postgraduate students in anthropology.

There is little disagreement though on the need to critically re-think anthropology's past and present insertion in the country's grossly unequal and still intensely racialised society, and to develop new approaches that reflect South Africa's postcolonial status. That said, there are vibrant, sometimes heated debates, as to where exactly the problems are located and what teaching anthropology and research in the discipline should aspire to. While agreeing on decolonisation as an indispensable response to colonialism and neo-colonialism, the underlying challenge is that decolonisation does not have a single meaning.

Regarding curriculum reform, of special relevance are debates as to whether changes should primarily entail "Africanisation", that is, the addition to, or possibly replacement of, the established canon of anthropology that has so far been taught mostly in the British social anthropology tradition with works by African and / or Black scholars.

A different perspective on decolonising knowledge production addresses the question of "How can social/cultural anthropology from the African continent help shape a new perspective on the world in the 21st century?" This perspective does not start from African-nationalist perceptions, but from listening to the voices that have emerged out of movements from below around the world, and broader questions about how decolonisation, critique and anthropology fit together.

This alternative approach to decolonisation calls for an intensive engagement with academic works as well as forms and ways of knowing that originate from different parts of the Global South, with the aim of challenging African-Western binaries. It thus goes beyond epistemological charges of "Africanisation" and the suggestion that Africans should conduct, exclusively, ethnographies of "their own people". It further emphasises the diversification of attention paid to forms of knowledge production, including a broad range of oral and written texts in order to include and acknowledge indigenous ways of knowing. The argument also presents challenges to what is deemed "anthropological"; it calls for writing across genres, including the boundaries between fiction and non-fiction, academic and popular so as to include novels, blogs, and performance.

The discussion of decolonised curricula in South African anthropology is furthermore opening up questions of decolonial pedagogy. Historically, the South African academy has emphasised that academic teaching equals the imparting of knowledge. In pedagogical practice, the lecture format has been prioritised, and authoritarian based expectations have been instilled in students. During current debates on curriculum reform, suggestions have been put forward to work against the authoritarian-colonial legacy by emphasising research- and writing-intensive teaching, including exploring and validating different forms of creative writing that go beyond or bypass the conventions of academic prose.

This pedagogy of the decolonial project has attracted some attention among anthropologists and resonates with the recent student movements' emphasis on disrupting the taken for granted. What does challenging established – authoritarian, colonial – assumptions about teaching mean for the complicated questions anthropology asks about the complexity, difference and diverse experiences of being human? How do content and pedagogy go together? In the remainder of the paper, I will present examples of decolonising epistemologies and curricula from four different South African universities.

Anthropological dispatches from four South African universities

As discussed elsewhere in this chapter, South African universities, including their Anthropology Departments, have distinct histories as spaces for the production of situated knowledges. This is a relevant concern of decolonisation, which may take on an array of meanings in different contexts and present diverse challenges in different locations due to divergent historical experiences. South African universities' pasts and presents differ substantially due to their institutional histories, their assigned places in the colonial and apartheid societies and the social backgrounds of students that they cater for in the post-apartheid society.

The differences between South Africa's universities are often neglected in debates about higher education and decolonisation. Often the experience of the formerly White, comparatively privileged institutions with British colonial roots, such as UCT, Rhodes and Wits, is assumed to reflect "the" South African academy. To a lesser extent searchlights have been shone onto institutions with an Afrikaner-nationalist history: Stellenbosch University, for instance, was for many years a battleground for the continuing supremacy of Afrikaans language and culture. In contrast, very little attention has been paid to the historically Black universities.

While the fall of Rhodes at UCT importantly kickstarted the decolonisation discourse in the South African academy, the critique of the formerly White, liberal universities in the British tradition needs to be understood as a particular history within South Africa and should not be transferred wholesale to a universalised conceptualisation of "the" South African academy, nor to a generalised perspective on the coloniality and decolonisation of anthropology. Rather, the divergent historical experiences have called for a range of responses. In the final section I will thus highlight a few examples of how anthropologists at different South African universities have responded to the local, national and international challenges of decolonisation.

Stellenbosch University: Revisiting race and politics

As discussed earlier, Stellenbosch University, near Cape Town, has been the elite university of Afrikanerdom. It was also the birthplace of *volkekunde* in the 1920s and 1930s. In the mid-2010s the university's history gave rise to an interesting focus on race and politics in the South African past and present.

In early 2013, Handri Walters, then a doctoral student in social anthropology, came upon a collection of human measuring instruments in the Stellenbosch University museum. This collection of "scientific" objects included Eugen Fischer's *Haarfarbentafel* (hair colour table), Rudolf Martin's *Augenfarbentafel* (eye colour table) and Felix von Luschan's *Hautfarbentafel* (skin colour table), as well as an anatomically prepared human skull. These objects had been used in studies of human measurement at Stellenbosch University (1925–1955) for the purposes of racial categorisation in support of racial and eugenic theories (often rooted in German academic literature), to inform constructions of the racial self and other. In her dissertation, Walters presented a detailed argument showing how eugenic theories had been employed "as vessels for revealing broader social, scientific and political narratives about race and racial classification, both historically" (Walters 2018). Objects such as those discovered by Walters were once widely used for teaching and research both in South Africa and internationally.

Walters' discovery at the Stellenbosch university museum inspired a collaborative project, which involved several of the Stellenbosch anthropologists. Named, *Indexing the Human: From Classification to a Critical Politics of Transformation*, the project ran between September 2014 and November 2015. *Indexing the Human* investigated the intersections of race, science, state policy and politics at Stellenbosch University in the past and connected those factors with race-thinking and racial science in post-apartheid South Africa. Funded by the Andrew W. Mellon Foundation, the project revolved around a weekly reading group, and 40 public events, including seminars, panel discussions and an exhibition.

Initially the project cast an eye on the German physical anthropologist Eugen Fischer, who had started his career in 1908 with a large-scale Southern African study on classification and racial mixing ("miscegenation"). He examined 310 children of the "Rehoboth Basters", a mixed-race community in German South West Africa (Namibia), applying the genetic theory of Gregor Mendel. His verdict was that African blood imparted impurity (Becker 2017). Fischer's *The Bastards of Rehoboth and the Problem of Miscegenation in Man* (1913) provided practical recommendations for German colonial policy. It can be read as a major colonial "pilot" of the racial science and eugenics research which he prominently led as the first Director of the Kaiser-Wilhelm-Institute for Anthropology, Human Heredity and Eugenics (KWI-A) in the 1920s and 1930s. The KWI-A played a key role as an institution of science in the implementation of Nazi racist ideology and policy during the holocaust. The Institute's story connects the Nazi atrocities of the 1940s with the colonial origins of racial science.

The Stellenbosch anthropologists employed Walters' encounter with a history of intellectual practice to investigate its "intimate proximity to our contemporary department and discipline" (Cousins and Reynolds 2016: 111). As Thomas Cousins and Lindsey Reynolds elaborated, the proximity of past and present was also "suggested by the distribution of a broader set of political economic conditions, processes, experiences and possibilities in which we currently live and work, that reveal continuities (and disjunctures) in how we think and act in contemporary South Africa" (Cousins and Reynolds 2016: 111). In an article published in the *Cape Times* newspaper at the time when the project started, the researchers stated an intellectual-political aim for and beyond South Africa – to "crack open some of our deepest held assumptions about what it means to be human", especially in the particular historical conditions of contemporary South Africa and the Global South (Cousins and Reynolds 2014)

University of the Free State: A seminar on decolonising knowledge

Stellenbosch University was the most elitist of the five formerly white Afrikaans-medium universities where anthropology was taught in the *volkekunde* tradition: Stellenbosch, the University of Pretoria, the University of the Free State (UFS), Potchefstroom University (after mergers, now known as Northwest University), and the Rand Afrikaanse Universiteit (RAU; now University of Johannesburg). The latter three institutions catered predominantly for the sons and daughters of rural Afrikaners. This section looks at an initiative by an anthropology lecturer at the University of the Free State, one of the "lesser" Afrikaner universities.

Albeit that the 2015–16 student protests highlighted the enduring colonial character of universities across South Africa, the legacies have taken different shape due to the different institutions' histories. One of the major characteristics of, and challenges for, the formerly Afrikaans universities was their long-time isolation from international and national academic discourses, especially the much debated power-knowledge nexus in the humanities. This issue was identified by Christian Williams, who had been appointed to a senior lectureship in anthropology at the UFS in 2014. His appointment was part of the UFS former Vice-Chancellor Jonathan Jansen's efforts to open up the formerly closed institution in Bloemfontein. Williams (2018: 53) argues that at the time of the student protests, the humanities at UFS had just begun "to emerge from this insular past". With the impetus of the student protests, thus, he initiated a seminar series titled *Decolonizing Knowledge* to open up transdisciplinary conversations about research and teaching at the university.

Throughout the 2017 academic year, academics and students met in weekly seminars under the headings "Unsettling Paradigms" and "Decolonizing Curriculum" for the joint discussion of significant texts by international key thinkers in postcolonial studies, including Frantz Fanon, Homi Bhabha, Walter Mignolo, Cheikh Anta Diop, and Kwame Appiah, as well as texts that discussed South African experiences of decolonising knowledge production. The investigation of South African instances started with the mid-1990s attempts by Mahmood Mamdani to reconstruct the African Studies curriculum at UCT, with a view to ending the "South African exceptionalism" paradigm in the academy. A counterpoint was set with a recent text by Mamdani (2016) titled *Post-Independence Initiatives in African Higher Education*. Other texts were even more locally-focused, such as an article by two North-West University anthropologists on institutional culture and public spaces on the Potchefstroom campus, another of the historically rural Afrikaans universities. Closest to home, the seminar agenda featured analyses of the infamous Reitz Affair at UFS, which had triggered a national and international outcry in 2008 when a group of white students posted an exceedingly racist and dehumanising video on social media (Williams 2018).

Williams believed that the best way to challenge the narrow-minded, and often enough racist, identity politics at the UFS was to open up debate by introducing anticolonial readings and the history of deconstructing colonial knowledge in an institutional context, "where questions of colonial knowledge and power have long been obscured" (Williams 2018: 84). His stance has been criticised as somewhat naive for believing in a tradition of postcolonial studies; such criticism has happened during exchanges at conferences and other public events, although I have yet to see a critique presented in writing. To some extent, his critics have challenged the notion of the exceptionality of the Afrikaans universities, the feature which Williams so strongly proposed.

The next example comes from UCT, as the most distinguished and oldest university in the British liberal tradition – supposedly the very opposite to the training grounds of rural Afrikanerdom.

UCT: "Making epistemic plurality possible"

The #RhodesMustFall uprising in March and April 2015 sent shockwaves through UCT, which had imagined itself for over a century in the liberal British tradition. This has been reflected in its institutional culture, set in a built-up environment complete with the ivy-clad buildings of the university's iconic upper campus, where, along with other humanities scholars, the social anthropologists have their offices.

Kharnita Mohamed joined UCT's Social Anthropology section in 2016; she found that the crucial moment had arrived, and an urgent call had emerged that, "after the first shockwaves of the #RMF and #FMF Movements, devising new curricula and rethinking pedagogy was urgent and possible" as she explained in a co-authored article on curriculum change in contested times (Rink et al. 2020: 10). Fallist student activists at UCT had been among the first to issue critical demands to revise the teaching and learning, assessment, administration and management practices at their institution. As student activists tirelessly pointed out, UCT's decidedly "English" atmosphere often left Black students feeling marginalised and voiceless. The institution responded with the establishment of the Curriculum Change Working Group (CCWG) in 2016, which facilitated debates about academic decolonisation and eventually presented a framework for curriculum change to the university (Rink et al. 2020: 3). Against this institutional background, curriculum reform became an urgent task for the institution's anthropologists too. Mohamed points out:

> "Whilst course redesign is not uncommon within anthropology and our department, student-led demands for decolonial pedagogies allowed epistemic and pedagogical shifts to happen." (ibid.)

Mohamed stated that her approach to decolonising the social anthropology curriculum was rooted in her Black feminist self-understanding "that recognised and grappled with our canons' androcentric and racist imaginaries" (Rink et al. 2020: 10). Yet she was also aware that anthropology had been a discipline that had been the space of "profound epistemic shifts borne by liberatory anthropological texts" (ibid.)

Put in charge of a third-year undergraduate course, "Anthropology through Ethnography", which had for many years formed a core component of the department's undergraduate curriculum, Mohamed defined as a key principle of decolonial teaching a desire to inspire plural epistemologies and "building epistemic communities, which encouraged multiple approaches to research questions" (Rink et al. 2020: 11). Practically, the 50+ students in the class were encouraged to express their own sense of liberatory anthropology and to form working groups based on their respective epistemic affinities. Working quite independently, the student teams produced blogs, reading lists and conducted research throughout the semester. Mohamed's lectures were aimed at facilitating collective enquiry-based learning rather than imparting knowledge. She argues that this pedagogical approach allowed students to appreciate the plurality of epistemologies and epistemic communities for knowledge production, thus challenging the discipline's established canons. She concludes that this gives room for a decolonial approach to academic learning, which she articulated as "giving students the tools to claim the right to produce knowledge because knowledge is not static, monolithic or eternal", and doing so "based on their concerns for the world, is where a decolonial curriculum resides" (Rink et al. 2020: 12).

Mohamed's response to her new teaching challenge thus ostensibly remained within the university's tradition of "liberal" plurality and "academic freedom" to opt for a range of analytical standpoints, yet it turned subtly subversive in its claim to an epistemic plurality that transcended the established and the canonical.

UWC: Decolonising the curriculum

In the final section I will focus on my own university and anthropology programme, where I have been teaching for the past twenty years. UWC was established in 1960 on the poverty-stricken periphery of Cape Town, some twenty kilometers from the city centre and UCT's glorious Table Mountain setting. The institution is a product of the 1959 apartheid act on higher education and was founded as an ethnic college for "coloured" (mixed-race) students. Today, its student body is overwhelmingly Black (in the inclusive sense, i.e., comprising "African", "Coloured" and a small number of "Indian" students), as well as admitting a small but increasing number of white students. Although the students' socioeconomic background is quite diverse, many students are graduates of under-resourced high schools in townships and rural areas and tend to be under-prepared for academic learning. Unlike formerly White universities, such as UCT or Stellenbosch, UWC has no endowment funds and legacies from wealthy graduates and industry to subsidise insufficient state funding. In addition, tuition fees at UWC are considerably lower than at the neighbouring "elite" universities of Cape Town and Stellenbosch. Consequently, class sizes are particularly large and resources for spaces that allow for experimental learning are scarce.

UWC proudly proclaims its history as a hotbed of the anti-apartheid struggle, known in the 1980s as the "intellectual home of the democratic left". Despite the institution's claim to rebellious "struggle" history, a peculiar authoritarianism has developed in the university's structures, becoming especially apparent during the 2015–16 uprisings. UWC's executive responded in a particularly unresponsive way to the demands of the student-worker struggles and the concerns of sympathetic academics (Becker 2019).

When protesting students burnt buildings, management shut down the campus for extended periods. The definition of what constitutes "violence" became a point of contention during the protests at many South African universities, but particularly so at UWC. In the spatial logic of apartheid violence, the university is located on the Cape Flats, an area of Black township settlements. The campus is fenced with only a few controlled entry points. Under these circumstances the student activists who continued to live in the on-campus residence halls found it hard to communicate with media, sympathisers from beyond the university, and even their own lecturers, who were for the most part barred from entering the campus during the revolt.

Tensions were rife between the older generation of academics and management, with roots in the university's history and its surrounding communities. Older UWC staffers proudly proclaimed their post-apartheid successes and condemned the young student protesters of 2015 and 2016 for being supposedly irresponsible. This was also reflected in an initially hesitant response to the call for decolonising the curriculum. With a few exceptions, the students' calls for decolonisation provided little immediate impetus to inspire academics to change their pedagogical approaches, and how they related to students or even their colleagues. As Bradley Rink, a human geographer and one of a small group of "concerned academics", which came together during the #FMF movements, writes, curriculum change happened in some pockets but overall, there was not much acceptance (Rink et al. 2020: 4).

The UWC anthropology curriculum had been changed quite profoundly around a decade earlier. Since about 2005 anthropology training at UWC had been modelled generally on the classical social anthropology undergraduate canon, with a few more contemporary add-ons; for instance, in the course on "Kinship and Gender" (previously known as "Kinship and the Family"), kinship studies were complemented with critical scholarship of gender, and addressed themes such as new reproductive technologies and "alternative" family forms.

Pedagogy continued to verge on the predictable, though, and teaching was often focused on conveying knowledge through lectures. Even so, the curriculum left some space for experimental and enquiry-based learning in individual courses, since lecturers were given the freedom to interpret the rather broadly-defined official syllabi ("course descriptors"). I, for one, had from the time I joined UWC, frequently included in-house individual and group research projects in my classes. I had also encouraged walking ethnography, a sensuous approach, and diverse genres of presentation, including creative writing and photography. Still, decolonisation was not an explicit aim in my courses until tentatively introduced in a half-course in 2017, and finally in the 2019 academic year when I presented a course with the explicit leading question: "How do we respond to the challenges of anthropology in postcolonial times?"

The Honours and Master's level students who were studying a core module in anthropological theory were encouraged to explore leading questions, including,

- What do ethnographies of Africa offer our efforts to understand the continent and the world at large?
- And, what may anthropology offer by way of crafting futures in Africa, and beyond?

The course thus moved away from its earlier format, which I had devised for teaching over many years; this had combined a first term of introducing and revisiting disciplinary concepts and epistemologies in chronological order with a second term focus on contemporary themes in anthropology.

Between 2006 and 2012, for instance, the theme was "glocality" and multiple modernities, which included topics drawn from the politics of culture and identity, as well as more theoretically oriented debates of "neoliberalism" and ethnographic investigations of responses to neoliberalism as a form of global capitalism. Between 2013 and 2015 the course design then responded to the affective turn. Pedagogically the course combined a textbook history and an intensive reading of original texts with a final inquiry-based essay, where students were encouraged to identify a topic of their interest within the given contemporary theme.

The 2019 version of the course focused in the first term on an in-depth intellectual and social history of South African anthropology in the 20th century, instead of the earlier broad sweep of "international" anthropology. The aim was to explore the diversity of the country's anthropologists, and hegemonic as well as marginal local disciplinary approaches. Pedagogically, this section of the course followed a rather conventional modus. The students were provided with an extensive reading list that included "classical" as well as recent and critical texts, including some from expressly decolonial perspectives, complemented by a series of guest lectures. For their research essay presentations, the students were asked to conduct extensive biographical and social context research on diverse South African anthropologists of the 20th century, including not only prominent white scholars such as Isaac Schapera, Monica Wilson, Philip Mayer and Werner Eiselen, but also a number of the Black South African anthropologists, including Z.K. Matthews, Livingstone Mqotsi, and Archie Mafeje. Lastly, it included anthropologists who are better known for their contributions in non-academic fields, such as the musician Jonathan (Johnny) Clegg, and activists David Webster and Mamphela Ramphele. The students were encouraged to select key works for discussion, including non-academic, artistic or activist contributions. In classroom discussions we considered the different anthropologists' significance for reinventing a decolonial South African anthropology.

The second term was dedicated to connecting different knowledge forms from the Global South with the students' southern African lived experiences (in addition to South-Africa raised participants, the class had students from Namibia, Zimbabwe, and Tanzania). The syllabus suggested this perspective in the form of "conversations" between scholars from the Caribbean and Africa about dynamics of colonialism, race, class and culture. We read core texts by C.L.R. James, Frantz Fanon, Walter Rodney, Stuart Hall and Rolph-Michel Trouillot.

The class was consistently urged to reflect on the reading and discussion of texts deriving from Caribbean backgrounds from the students' own autoethnographic perspective. The guiding question was, "How do they speak to our southern African experience?"

At first, the students found it hard to relate to the intense reading and critical discussion of African scholarship and the writings by authors from the Caribbean, who were, with the exception of Fanon, entirely unknown to most of them prior to taking this course. Students struggled with the texts, partly because they had no knowledge of Caribbean history, and thus initially found it hard to see possible connections and parallels with southern African conditions. They also articulated that while they were keen to engage with the themes of race and colonialism, they were reluctant to delve into other analytical approaches, particularly Rodney's focus on class analysis and capitalism as the material basis of racism. Interestingly, several of the students eventually became quite intrigued by "bringing class back in". One of the students, for instance, expressed her understanding in a (well-phrased) response to the pedagogical process:

> "Rodney's analysis of racism cannot be separated from the inequality created by a capitalist economic system. But this does not mean that white supremacy did not evolve into its own cultural phenomena. … In conclusion, Rodney's economic analysis of colonialism and race is still very relevant today because Africa is still very much affected by the consequences of the exploitation and theft of resources that historically took place at the hands of Europe. Africa's underdevelopment and poverty is directly related to slavery and the colonial system, which is an economically exploitative system rather than a racist system. Racism was a by-product of the system and not its cause."

Classroom conversations were lively, quite in contrast to earlier experiences with teaching "theory", where proceedings had in the past often been rather dreary. In contrast, students of the 2019 cohort began to voice that, while most of them navigate multiple life worlds every day, they find it difficult to articulate questions about these (see also White 2019). They spoke with increasing urgency about their concerns that for most people "out there" (in their "communities") social difference is ontological, and that they have found it hard to respond to these sentiments from the background of anthropological perspectives on difference, particularly matters of race and their intersection with other categories of difference, such as class and gender.

The students found the seesaw of auto-ethnography and thought by scholars from a comparable, yet different context of the Global South helpful to sift through the trajectories of anthropology and its paradoxical relationship with domination and resistance in South Africa. The encounter seemingly contributed to the understanding that while anthropology has historically been entangled with colonialism, anthropologists have also been questioning the orthodox by asking questions about being human in different contexts. In other words, students arrived at an understanding that while anthropologists have been complicit in the making of dominant ideologies, they have nevertheless played a central role in critiquing hegemonic discourses and unravelling their effects.

These notes provide an outline of individual efforts of reinterpreting existing curricula. In 2019/20 our newly autonomous Anthropology Department (between 1995 and 2018 UWC's anthropology and sociology programmes had been housed in a merged department) embarked on a collective effort of curriculum reform. While we streamlined the programme and decreased the number of courses taught, we introduced new thematic undergraduate courses on cultures of capitalism, and environmental anthropology respectively. However, the most radical approach we took concerned first year teaching and learning, in regard to both content and pedagogy.

We particularly decided to avoid teaching canonical introductions to anthropology, which typically revolve around theoretical and methodological concepts such as "cultural relativism" and "participant observation", taught with reference to key "international" anthropologists from the (Anglo-Saxon) Global North.

Instead, since 2021 our first-year students have started their anthropological training with an enquiry-drawn and writing intensive exploration of the discipline's key question: What does it mean to be human?[9] The first semester course explores the question of the human through a series of themes each of which highlights a particular set of debates and enquiries about the making of the category "human". The course invites students to explore the following questions:

9 My colleagues Kelly Gillespie and William Ellis were central to the design of the first semester course, while I took the lead in developing the second semester course.

- What does it mean to be human?
- How has this question been answered in different historical eras and in different places?
- Who / what is excluded from the human?

The course begins with different interpretations of how humans came into being, "origin stories" from different eras and world regions; an initial assignment invites students to explore the origin story they have grown up with in their family. The course follows different moments in the history of the human from the emergence of human hands through to the globalisation of slavery and the invention of the human as commodity. It concludes with the recent turn in thinking about humans as post-human. Students thus learn how "being human" is not a natural fact, but a complex historical, social and political process that is constantly being debated and challenged. Before its closing exploration of the Anthropocene, the course introduces the students to questions of race and the human, the hierarchising of species and the role of colonial violence and the museum, as well as debates about strategies of decolonial rehumanisation.

In the second semester the course revolves around representations of difference and explores shifting notions of inclusion and exclusion in various historical and local contexts, and how those challenge, and are in turn challenged by the decolonial project. The semester starts with asking the students to deliberate on the concept of beauty with which they have grown up and how this contributes to categories of indexing humans, such as race, gender and class. We then relate multiple histories of human difference. These include, human disposability, intersections of violence in global history (with a focus on the German genocide in Namibia and the holocaust), racial capitalism, and the gendered dimensions of violent histories of difference.

Through the themes explored in both semesters the students are introduced to the questions which anthropologists grapple with in enquiries of race and the human, the role of colonial and capitalist violence, representations of difference, shifting notions of inclusion and exclusion, and issues of social (in)justice. These efforts to decolonise the anthropology curriculum explicitly respond to the South African key challenges of inequality and social justice.

The pedagogical approach that I describe emphasises critical inquiry through intensive writing practice.[10] Through pedagogical practice and the questions that the course asks, my colleagues and I strive to create space for students to speak and to listen to each other's experiences of navigating multiple life worlds every day. The emphasis on creating space for listening encourages peer learning; listening is also critical for anthropologists teaching in South African universities to learn how to imagine how decolonisation, critique and anthropology fit together. This is particularly important in an academic institution where the life worlds of middle-class academics and working-class students diverge as sharply as they do in historically Black universities, such as UWC.

Conclusion

This essay started with a broad sweep of the frequently contested historical development of South African anthropology. I demonstrated how anthropology in South Africa, from its inception a century ago, has been implicated in the country's political-economic history at every turn – from the "native question" of the colonial era through to the discipline's uses and abuses during apartheid, and also its counter-hegemonic traces. In post-apartheid South Africa anthropology has continued to be shaped by socio-economic and cultural processes. The discussion then turned to the challenges and contestations that South African anthropologists are currently navigating in response to the forceful calls by student-led movements for decolonisation in their still intensely racialised and socially unequal society. By putting the efforts of decolonising South African anthropology into context, it has become clear not only that anthropologists at different South African universities have devised a broad range of self-reflective and dynamic strategies in decolonising the discipline, but also shown how their strategies respond to their respective institutional contexts. It furthermore shows how different contexts of coloniality in the Global South and North call for different strategies for decolonising the curriculum, the academy and society at large. And finally, in

10 Kelly Gillespie brought the Writing Intensive Teaching (WIT) approach to the UWC Anthropology Department and invited Pamela Nichols, the head of the Wits Writing Programme, for hands-on training during a series of online and in-person workshops with the department's teaching and tutoring staff.

pointing out the significance of the South African student movements that also inspired decolonisation and epistemic disobedience in the Global North, my account shows that decolonisation is now a truly transnational movement.

Bibliography

Adebanwi, W. (2017): "Elites, Ethnographic Encounters and the 'Native' Ethnographer in Contemporary Africa." In: R. Boswell/F. Nyamnjoh (eds.), Postcolonial African Anthropologies, Cape Town: HSRC Press, pp 149–76.

Ahmed, A. K. (2019): "#RhodesMustFall: How a Decolonial Student Movement in the Global South Inspired Epistemic Disobedience at the University of Oxford." In: African Studies Review 2019, pp. 1–23.

Bank, A. (2015): "'Broederbande' [brotherly bonds]: Afrikaner nationalist masculinity and African sexuality in the writings of Werner Eiselen's students, Stellenbosch University, 1930–1936." In: Anthropology Southern Africa 38/3-4, pp. 180–97. https://doi.org/10.1080/23323256.2015.1069715.

Bank, A. (2016): Pioneers of the Field: South Africa's Women Anthropologists, Johannesburg: Wits University Press. https://doi.org/10.1017/CBO97813165 84187.

Bank, A./Bank, L. (eds.) (2013): Inside African Anthropology: Monica Wilson and Her Interpreters, Cambridge: Cambridge University Press. https://do i.org/10.1017/CBO9781139333634.

Bank, L. (2011): Home Spaces, Street Styles: Contesting Power and Identity in a South African City, Johannesburg and London: Wits University Press and Pluto.

Becker, H. (2012): "Anthropology and the Study of Popular Culture: A Perspective from the Southern Tip of Africa." In: Research in African Literatures 43/4, pp. 17–37. https://doi.org/10.2979/reseafrilite.43.4.17.

Becker, H. (2017): "Auschwitz to Rwanda: the link between science, colonialism and genocide." In: The Conversation, January 26, 2017. (https://theconvers ation.com/auschwitz-to-rwanda-the-link-between-science-colonialism-and-genocide-71730)

Becker, H. (2018): "Dissent, Disruption, Decolonization: South African Student Protests, 1968 to 2016." In: International Socialist Review 111, pp. 31–47.

Becker, H. (2019): "Student-Worker Struggles for Fair Labour Practices on University Campuses", Rosa Luxemburg Foundation Southern Africa. (htt ps://rosalux.co.za/publication/student-worker-struggles-for-fair-labour =practices-on-university-campuses).

Boonzaier, E./Sharp, J. (eds.) (1988): South African Keywords: The Uses and Abuses of Political Concepts, Cape Town: David Philip.

Boswell, R. (2017): "Introduction: Continuities and contradictions in postcolonial African anthropologies." In: R. Boswell/F. Nyamnjoh (eds.), Cape Town: HSRC Press, pp 1–12.

Coertze, P. J. (1959): Inleiding tot die Algemene Volkekunde [Introduction to General Volkekunde], Johannesburg: Voortrekkerpers.

Cousins, T./Reynolds, L. (2014): "We are all in the same boat now in SA." Cape Times, September 12, 2014. (https://www.iol.co.za/capetimes/opinion/we -are-all-in-the-same-boat-now-in-sa-1750344).

Cousins, T./Reynolds, L. (2016): "Indexing the human: from classification to a critical politics of transformation." In: Anthropology Southern Africa 39/2, pp. 110–115. https://doi.org/10.1080/23323256.2016.1174071.

Fischer, E. (1913): "Die Rehobother Bastards und das Bastardierungsproblem beim Menschen: anthropologische und ethnographiesche Studien am Rehobother Bastardvolk in Deutsch-Südwest-Afrika, ausgeführt mit Unterstützung der Kgl. preuss, Akademie der Wissenschaften." In: Z.Vererbungslehre 10, pp. 280. https://doi.org/10.1007/BF01943440.

Gluckman, M. (1940): "Analysis of a Social Situation in Modern Zululand. ('The Bridge')." In: Bantu Studies 14/2, pp. 147–74. https://doi.org/10.1080/02561 751.1940.9676107.

Gluckman, M. (1960): "Tribalism in Modern British Central Africa." In: Cahiers d'Études Africaines 1, pp. 55–70. https://doi.org/10.3406/cea.1960.2939.

Gordon, R. J. (2018): The Enigma of Max Gluckman: The Ethnographic Life of a "Luckyman" in Africa, Lincoln and London: University of Nebraska Press. h ttps://doi.org/10.2307/j.ctv47w743.

Gordon, R./Spiegel, A. (1993): "South Africa Revisited." In: Annual Review of Anthropology 22, pp. 83–105. https://doi.org/10.1146/annurev.an.22.10019 3.000503.

Hartnack, A. (2013): "On Gaining Access: A Response to Francis Nyamnjoh's 'Blinded by Sight: Divining the Future of Anthropology in Africa." In: Africa Spectrum 48/1, 107–112. https://doi.org/10.1177/000203971304800106.

Hellmann, E. (1948): Rooiyard: A Sociological Study of an Urban Native Slumyard, Cape Town: Oxford University Press for Rhodes-Livingstone Institute.

Hunter, M. (1936): Reaction to Conquest: Effects of Contact with Europeans on the Pondo of South Africa, London: Oxford University Press.

James, D. (1997): "Anthropology, History, and the Making of Past and Place." In: P. McAllister (ed.), Culture and the Commonplace. Anthropological Essays in Honour of David Hammond-Tooke, Johannesburg: Witwatersrand University Press, pp. 115–36. https://doi.org/10.1080/00020189708707871.

Le Grange, L. (2019): "The Curriculum Case for Decolonisation." In: J. D. Jansen (ed.), Decolonisation in Universities. The Politics of Knowledge, Johannesburg: Wits University Press, pp. 29–47.

Maldonado-Torres, N. (2016): Outline of Ten Theses on Coloniality and Decoloniality. Frantz Fanon Foundation (online) https://fondation-frantzfanon.com/outline-of-ten-theses-on-coloniality-and-decoloniality/ (Accessed 17 May 2024)

Mamdani, M. (2016): "Between the Public Intellectual and the Scholar: Decolo nization and some Post-independence Initiatives in African Higher Educ ation." In: Inter-Asia Cultural Studies 17/1, pp.68-83. https://doi.org/10.1080/14649373.2016.1140260.

Mamdani, M. (2019): "Decolonising Universities." In: J. D. Jansen (ed.), Decolonisation in Universities. The Politics of Knowledge, Johannesburg: Wits University Press, pp. 15–28. https://doi.org/10.18772/22019083351.6.

Matthews, Z.K. with a Memoir by M. Wilson (1981): The Autobiography of Z.K. Matthews: southern Africa 1901 to 1968, London: Rex Collings in association with David Philip (Cape Town).

Mayer, P./ Mayer, I. (1971): Townsmen or Tribesmen: Conservativism and the Process of Urbanization in a South African City, Cape Town: Oxford University Press.

Mqotsi, L. (1957): "A study of *ukuthwasa*: (being a syndrome recognized by the Xhosa as a qualification for being initiated as a doctor)". MA thesis, University of the Witwatersrand (published in 2007 as a book, East London: Iqula Press).

Mqotsi, L. (1989): The House of Bondage, Johannesburg: Skotaville.

Naidoo, L.-A. (2016): "Contemporary Student Politics in South Africa': The Rise of the Black-led Student Movements of #RhodesMustFall and #FeesMustFall in 2015." In: Anne Heffernan/Noor Nieftagodien (eds.), Students Must Rise: Youth Struggle in South Africa Before and Beyond Soweto '76, Johannesburg: Wits University Press, pp. 180–190. https://doi.org/10.18772/22016069193.24.

Ndlovu-Gatsheni, S. (2018): Epistemic Freedom in Africa: Deprovincialization and Decolonisation , London and New York: Routledge. https://doi.org/10.4324/9780429492204.

Niehaus, I. (2013): "Anthropology and Whites in South Africa: Response to an Unreasonable Critique." In: Africa Spectrum 48/1, pp. 117–27. https://doi.org/10.1177/000203971304800108.

Nyamnjoh, F. B. (2012): "Blinded by Sight: Divining the Future of Anthropology in Africa." In: Africa Spectrum 47/2–3, pp. 63–92. https://doi.org/10.1177/000203971204702-304.

Rink, B./Mohamed, K./Ngoasheng, A./Behari-Leak, K. (2020): "Crossing Borders as 'new' Academics in Contested Times: Reflexive Narratives of Curriculum Change and Transformation." In: Critical Studies in Teaching & Learning 8/1, pp. 1–20. https://doi.org/10.14426/cristal.v8i1.229.

Salo, E. R. (2018): Respectable Mothers, Tough Men and Good Daughters: Producing Persons in Manenberg Township South Africa, Mankon, Bamenda: Langaa Research & Publishing. https://doi.org/10.2307/j.ctvh9vz0q.

Sharp, J. S. (1981): "The Roots and Development of Volkekunde in South Africa." In: Journal of Southern African Studies 8/1, pp. 16–36. https://doi.org/10.1080/03057078108708032.

Spiegel, A./Becker, H. (2015): "South Africa: Anthropology or Anthropologies?" In: American Anthropologist 117/4, pp. 754–762. https://doi.org/10.1111/aman.12353.

Spiegel, A. (2005): "From Exposé to Care. Preliminary Thoughts About Shifting the Ethical Concerns of South African Social Anthropology." In: Anthropology Southern Africa 28/3-4, pp. 133–41. https://doi.org/10.1080/23323256.2005.11499923.

Spiegel, A. D./Becker, H. (2018): "South Africa, Anthropology." In: Hilary Callan (ed.), International Encyclopedia of Anthropology, New York: John Wiley & Sons.

Van der Waal, C. S. (Kees) (2015): "Long Walk from *Volkekunde* to Anthropology: Reflections on Representing the Human in South Africa." In: Anthropology Southern Africa 38/3–4, pp. 216–34. https://doi.org/10.1080/23323256.2015.1088392.

Van Wyk, I. (2012): "A Response." In: Anthropology Southern Africa 35/3–4, pp. 119–21. https://doi.org/10.1080/23323256.2012.11500035.

Walters, H. (2018): Tracing Objects of Measurement: Locating Intersections of Race, Science and Politics at Stellenbosch University, Stellenbosch: Stellenbosch University (PhD in Social Anthropology).

White, H. (2019): "What is Anthropology that Decolonising Scholarship Should be Mindful of?" In: Anthropology Southern Africa 42/2, pp. 149–160. https://doi.org/10.1080/23323256.2019.1574210.

Williams, C. A. (2018): "Decolonising Knowledge: Reflections on Colonial Anthropology — a Humanities Seminar at the University of the Free State". In: Strategic Review for Southern Africa 40/1, pp. 82–103. https://doi.org/10.35293/srsa.v40i1.274.

Wilson, G. (1939): The Study of African Society, Livingstone, Northern Rhodesia: Rhodes-Livingstone-Institute.

Wilson, M./Mafeje, A. (1963): Langa: A Study of Social Groups in an African Township, Cape Town: Oxford University Press.

(Re)Building Epistemology or (Re)Shaping Societal Outlook

A critique of the Sudan's Islamisation of knowledge paradigm

Bakheit Mohammed Nur

Introduction

Across Africa, over the course of the 20th century, educationists, policymakers, intellectuals, and theologians began to turn their focus toward the subject of Islamic higher education on the continent. The nascent Islam-informed field of epistemological development, and its potential ability to serve as a response to contemporary Muslim societal challenges, began to dominate scholarly debate. In the 1970s, these intellectual pursuits led to the emergence of a new movement called "the Islamisation of knowledge". This new epistemic paradigm aimed – and still aims – to (re)shape higher education policy and practice in accordance with Islamic teachings. It seeks to combine academic excellence with Islamic knowledge, incorporate Islamic teachings and ethics into the university curriculum, and (re)map the political economy of knowledge production in African Muslim societies. Proponents of this paradigm seek to promote Islamic epistemology through organising academic conferences, seminars, symposia, and lectures on the challenges facing Muslim higher education.

Participants in these forums recognise that the state of Muslim higher education in Africa and elsewhere is tenuous, and they envision possibilities for its pedagogical reformation and transformation. These measures of reconfiguration give rise to Islamic higher educational consciousness among Muslims and heighten the urgent need for modern Islamic universities, colleges, research institutions, and centres that approach knowledge from an Islamic epistemology perspective. Since the emergence of the Islamisation of knowledge paradigm, Islamic universities and colleges have been established in countries across Africa, even in minority Muslim countries such as Kenya, Tanzania, Burkina Faso, Mozambique, South Africa, and Uganda, to provide tertiary education and skilful training in various fields of Islamic, social, and natural sciences (Lo and Haron 2016). Furthermore, institutions and individuals have established publication houses and academic journals across African Muslim societies to decolonise knowledge, develop Islam-oriented educational materials, and spread new epistemic theories and methodologies responsive to Muslim worldviews and ontological realities.

However, the Islamisation of knowledge project seems to suffer serious epistemological and methodological shortcomings, rendering its reconfiguration efforts ineffective in dealing with issues of coloniality of power, asymmetry in knowledge production, and Muslim educational challenges. Thus, rather than producing an integrated epistemology to advance cognitive emancipation, the project's educational practices on the ground instead perpetuate the educational dichotomies (Western vs. Islamic) created during European colonialism. This article discusses some of these methodological shortcomings and epistemological contradictions. It highlights the socio-cultural diversity within African Muslim societies and epistemological differences in knowledge production between Islamic and Western paradigms. This research explores how the Islamisation trend attempts to eschew Western-centred knowledge and refine or (re)define modern disciplinary specialisations according to an Islamic epistemology perspective. It also engages a broader discourse on Muslim knowledge politics, critically highlighting various dichotomisation narratives such as reason vs. faith, epistemic certainty vs. scepticism, dualism vs. integration, and androcentric values vs. liberation. The article enriches and substantiates the discussion on these various thematic subjects with a thick ethnography and rigorous anthropological analysis drawing from field research in the Sudan in 2019 and 2020. It mediates the diverse voices of protagonists and antagonists in the domains of Islamic epistemology in general and the Islamisation of knowledge in particular.

Islamisation and the politics of its knowledge (re)production and transmission are complex, multi-dimensional epistemic phenomena with serious institutional repercussions in the Sudan. The unfolding discourse on the Islamisation of knowledge in the Sudan between 1989 and 2019 intertwined with the construction of a new Islamic state imagined and promoted by Islamist intellectuals who were inspired by a narrow interpretation of Islamic law to establish political legitimacy and governance. This (re)Islamisation of society, knowledge, economics, politics, media, institutions, and public life led to the spread of religious symbols and activities in various state apparatuses. With the inception of an Islamist regime in 1989, the Islamisation of knowledge paradigm became a state-sponsored epistemic project, emerging within the matrix of a reinforced Islamic resurgence that attempted to combine modernity with faith. As part of my anthropological investigation of this phenomenon, I employed various social research methods to comprehend the theoretical dimensions of this subject, as well as its practices and methodologies on the ground. Between 2019 and 2020, I conducted six months of intermittent but intensive ethnographic research in Khartoum State. I chose this site for research because it is home to the Sudan's three major Islamic universities: Omdurman Islamic University, the University of Holy Qur'an and Islamic Sciences, and the International University of Africa. Khartoum State is significant in the context of the Islamisation trend and for its operationalisation strategies at institutional level. It also embraces religious groups with opposing epistemic orientations, ideologies, philosophies, learning styles, and intellectual traditions. Khartoum State is home to the Sudan's capital city of Khartoum, where education policy is created and dispatched throughout the country and beyond. Furthermore, Khartoum State has deep roots in traditional higher Islamic learning, characterised by both embodied and intellectual approaches to Islamic knowledge production and transmission, as well as by secular-oriented academic institutions of higher education that were established during British colonialism in the early 20th century.

Part of my fieldwork consisted of conducting interviews with key players in this field. I interviewed policymakers who chart the course of higher education in the Sudan and university administrators who implement these policies. I interviewed the educators and lecturers who impart Islam-oriented knowledge and practice to their students. In these encounters, I sought to learn more about the academic backgrounds of these educators and hear their own reflections on the curricula that they prepare. I interviewed students who receive this knowledge and its representations, aiming to learn more about their own reading and perceptions of the religious textbooks and other educational materials they use in their programmes. We also discussed issues such as enrolment in Islamic universities and the motives behind those decisions. I engaged with scholars who work on the Islamisation of knowledge within Islamic universities or research centres and who either promote or oppose specific epistemic orientations and ideologies.

To enrich my ethnography, I conducted intimate focus group discussions with small groups of interlocutors including research participants with shared experiences, expertise, and practices. Groups included, for example, cohorts of students, lecturers, policymakers, educators, pedagogists, and administrators. I considered gender, age range, and social background in forming these groups to ensure their productivity. I guided discussions by asking open-ended questions that would elicit insight into participants' attitudes, beliefs, opinions, perceptions, and practices surrounding Islamisation policy and procedures. Through group discussions, I sought to bring together members of a shared practice to articulate and reflect in friendly conversations about their understandings of and experiences with the Islamisation process and its related practices within the three Islamic universities. The advantage of this method was that it allowed interlocutors to interact with, validate, and question each other on these topics. Furthermore, it generated a social environment in which conversations occurred more naturally than they might have in individual interviews. Such group discussions, in turn, assisted me in exploring new ideas and unforeseen issues regarding the Islamisation of knowledge and helped me clarify opinions that were held by members of a group. They also led me to other potential interlocutors considered to be knowledgeable or active in this domain. In these discussions, I paid attention to nonverbal communication cues and reactions, such as facial expressions, emotions, and tones of voice.

During group discussions, I not only paid attention to what interlocutors were saying, but also made inferences based on their actions. I attended lectures and seminars with students to observe the interactions between lecturers and students and to learn more about the ideas that were communicated in those milieux. My participation in the lectures and seminars also offered insight into how learning and knowing are created and negotiated inside the classroom. Outside the classroom, I socialised with students to become more familiar with their educational perspectives. Utilising a similar technique with lecturers enabled me to obtain valuable ethnographic data by revealing discrepancies between what interlocutors were saying and what they were doing in their day-to-day activities within and outside universities. I developed rapports with research participants to consolidate data collection. I also attended workshops and conferences organised by Islamic universities and their partners to uncover the scholarly networks and research interests these universities pursue. I participated in public lectures organised by Islamic universities on campuses and elsewhere. These lectures formed scholarly forums where students, lecturers, university administrators, and the general public exchanged and debated ideas. Through my participation in public lectures, I discerned which ideas, thematic issues, and forms of knowledge Islamic universities intended to transmit to the public and why. This activity also afforded me an opportunity to interview the organisers, speakers, and audience members.

The discussion in this article is not a comprehensive treatment of the subject of Islamisation of knowledge, but rather, a brief articulation of some of its shortcomings and epistemological contradictions, evident in the methodology and practices of the Islamisation project. These contradictions include, for instance, diversity vs. homogeneity discourse, refining vs. creating disciplines, scepticism vs. certainty, reason vs. faith, dichotomous vs. integrative curricula, and conservatism vs. liberalism.

Diversity vs. homogeneity

The practice and experience of Islam in one culture or society may appear to be distinct from those in others. Muslim societies hold a variety of beliefs, practices, ideologies, epistemologies, methodologies of learning, social structures, and cultural values. Manger (2002) argues that this diversity is sustained by a broad spectrum of Islamic experience, knowledge systems, and educational philosophies. The textual sources that Muslims most frequently turn to in their learning institutions are the Qur'an, the *ḥadîth*, *ijma'* (the juristic consensus), and *qiyâs* (reasoning by analogy). These sources also form epistemic principles of knowledge production and transmission that supersede theological realms and hermeneutics. However, Muslims in any given society consult, study, and interpret these scriptural texts differently, depending on their own socio-cultural and/or ideological justifications and logics. On Muslim diversity, Sardar argues that "we Muslims live as composite identities: various interpretations of Islam, with their various histories, cultural, regional, ethnic variations, and now national divergences, are always part of our individual and collective personalities" (Sardar 2004: 184). It is common for Muslims who come from different cultural backgrounds and nationalities and who possess diverse worldviews and philosophies to interpret Islamic texts and sources differently. It is also normal for Muslims to use those texts in different ways and for different purposes, depending on their understandings and the ideological or epistemological principles on which their interpretations are based. This variety in

Muslim practice impacts the Islamisation of knowledge paradigm and its epistemic orientation in the Muslim world today.

Multiple interpretations and different forms of knowledge exist not only among Muslim scholars but also among ordinary Muslims who are influenced by or exposed to Islamic scholarships via prevalent discourse on the politics of knowledge production and transmission. Such plurality manifests itself in different societal aspects, such as gender, class, ethnicity, education, and interpretation of knowledge. This diversity strengthens the notion that Islam comprises not only cosmological themes of the holy texts, but also lived identities, ideological orientations, epistemological approaches to knowledge, and experiences in social contexts, each emerging within ongoing debates and discourses about what is right and what is wrong when viewed from socio-cultural and religious perspectives (Manger 2013: 18). Such a multiplicity is manifested in education, which encompasses a wide spectrum of learning experiences, methods, forms of knowledge, and scholarly practices across Muslim societies.

The call to shift epistemic orientation from Western to Islamic is a common theme in the discourse on decolonisation and Islamic educational reform. Because they have initiated and proceeded to implement its policy and practices on the ground, Islamisation of knowledge enthusiasts in general and reformation activists, such as Islamists attempting to build an Islamic state in particular, have the louder [or stronger] voice in this epistemic reformation process. They have politicised the Islamisation of knowledge discourse since its inception, incorporating its epistemic narratives into their ideological agenda. This concern [endeavour] has also drifted into epistemological debates over de-Westernisation, de-secularisation, de-colonisation, and educational reform directed toward political activism imbued with Islamism tendencies and its version of Islamic interpretation. Thus, the movement to Islamise knowledge pays little attention to the diversified views held by Muslim scholars and intellectuals about the nature of knowledge and knowing.

The movement constructs its conceptualisation of educational problems and epistemic future prospects on homogenous and universalistic categorisations such as the *ummah* as a way to address Muslims collectively and unify their epistemic dilemma. Here, the *ummah* is a universal community of believers whose members commit themselves to Islam and its socio-religious orders. Beneath the surface of a common commitment to Islam, there are variety of ethnicities, cultures, socio-cultural orders, and intellectual traditions that cannot be overlooked when designing a new epistemic policy and practice to be pursued and implemented by all Muslims. Islamisation theorists use the term *ummah* in its original form and insist that it should not be translated into other languages lest its Islamic meaning is distorted (Al-Alwani 2005; Al-Fârûqî 1989; Sardar 2004). In his critique of the Islamisation project and its pioneers, Sardar explains that the religious commonality established through the concept of *ummah* "is not synonymous with 'the people', or 'the nation', or 'the state' or any other expression which is determined by race, geography, language, history or any combination of these" (Sardar 2004: 194–195). The term, therefore, becomes a methodological tool for transcending Western political concepts that confine individuals into certain identifications and territorial belongings. Protagonists of the movement try alternatively to establish faith-based universalism that transcends locality, regionalism, nationalism and other forms of limited belonging. They communicate the notion that the concept is trans-geographical, trans-local, trans-cultural, trans-racial, and trans-class. This implies that the concept is non-discriminatory in the context of Muslims and thus it can be utilised analytically for the development of unitary Islamic epistemology and its implementation thereafter. The concept of *ummah* therefore serves as a referent of unity and cooperation and shapes the historical and contemporary consciousness of Muslimness as a common identity that surpasses socio-political and cultural boundaries and pluralism. The notion of *ummah* goes beyond the politics of homeland and nation-state intending to offer an alternative conception of universality built ontologically from the Islamic ethos and divine imperative. It influences unitary consciousness, shapes political behaviour of solidarity, and creates a conception of international society whose commonality and collectiveness are derived from Islam-centred civilisation, epistemology, and way of life (see Shani 2008). In this discourse, educational policy that draws its inspiration from sectarianism, Sufism, or traditionalism is regarded as divisive and anachronistic and therefore inappropriate or irrelevant to the projects of decolonisation and Muslim educational reform.

It is for this reason that the movement to Islamise knowledge overlooks Sufi and traditionalist approaches to knowledge production, claiming that they are backward, repetitive, and limited to theology, rote memorisation, and asceticism and therefore cannot deal with contemporary Muslim societal and educational problems. This limitation meets resistance to the Islamisation of knowledge programmes in Muslim societies where Sufism and traditionalism are dominant religious forces on the socio-political and educational landscape. The resistance often regards the Islamisation of knowledge as ideological Islamist training rather than as an epistemological reform initiative intended to liberate Muslim minds from coloniality and advance Muslim societies. Mabud argues that the need to make education Islamic is perhaps shared by many Muslim scholars. However, the definition of Islamisation, its methodological approach and epistemological contents are greatly contested, as these scholars come from different intellectual traditions and cultural backgrounds (Mabud 2016: 132). Thus, diverse Islamic philosophical understandings of knowledge yield multiple scholarly approaches to the idea of the Islamisation of knowledge.

Islamisation of knowledge proponents treat Muslim communities in isolation, as if the detrimental influence of Western epistemology is experienced only by Muslims or as if the epistemic problem of other religious communities living with Muslims is not their main educational concern. Thus, the Islamisation of knowledge movement pays little or no attention to the presence of non-Muslim populations who live in Muslim majority countries (Ashraf 1990). This indifference to other faiths' educational premises becomes problematic in national educational discourse in some countries where Christians and other religious communities constitute a considerable number. For instance, when Islamists introduced the Islamisation of knowledge and society in the Sudan, people from southern part of the country, the Blue Nile Region, and the Nuba Mountains, regarded it as an internal neo-colonialism since its programmes were designed to (re)shape the epistemic and socio-cultural values of all Sudanese people in accordance with Islamic and Arabic culture. The reason behind this conception was that the Islamisation of knowledge movement has failed to produce comprehensive epistemic reform programmes that are acceptable to all cultures within Sudanese society.

The Islamisation of knowledge project becomes more problematic when considering the management of diversity in the context of pluralistic and multi-cultural society. The movement calls for a unified system of education in which secular and Islamic schoolings are integrated into one unitary educational scheme. In this union, Islam is considered to be the driving spiritual force behind educational policy and practices in elementary and secondary schools as well as in colleges and universities. The epistemological union is believed to eliminate existing shortcomings, such as so-called archaic textbooks, untrained teachers at traditional Islamic schools, and the alienating mimicry of the secular West. Al-Fârûqî (1989) claims that the condition of being civil consists in the acquisition of knowledge of Islamic civilisation, which covers subjects such as the ethics, law, culture, and heritages of Islam. He argues for mandatory acquisition of this knowledge for Muslims and non-Muslims alike at university level, regardless of their specialisations and religious professions. He states:

> "It is not possible to be civil without such knowledge [of Islamic civilization]. Even if students belong to a non-Muslim minority, it should not absolve them from fulfilling this basic requirement. Since they or their parents have opted to become citizens of the Islamic state, they must acquire the necessary familiarity with the civilization in which they are living and with the spirit and hope that move this civilization and its compatriots. No person may be left without such acculturation, which is basic for "socialization" or integration into Muslim society" (Al-Fârûqî 1989: 16).

Al-Fârûqî condemns colonialism for imposing Western culture and epistemology onto Muslim societies. In his education philosophy, however, he proposes an epistemic policy that is very similar to the colonial one when he makes Islamic education mandatory for minority non-Muslims who live among Muslims. These minorities, without doubt, have their own ideas about what kind of education they want to pursue. The very idea of demanding non-Muslims' familiarity with Islamic civilisation seems to contradict Al-Fârûqî's emancipatory arguments.

In order to address the shortcoming of this epistemic philosophy, the Islamic Academy of the United Kingdom organised a series of public lectures, seminars, and conferences at the University of Cambridge in 1989 and 1990 (see Mabud 2016). These forums brought together Muslims from different intellectual traditions and Christian educationists to establish a dialogue and discuss faith as a shared epistemological basis for educational reform that would draw from the spiritual values of both Islam and Christianity. The meetings sought to find a common spiritual ground for faith-based education in a multi-faith and multi-cultural society, aiming at establishing a shared set of beliefs held by Muslims, Christians, and other religious believers. The unifying epistemological characters of Islam and Christianity centre on the belief in the necessity for divine guidance, transcendental afterlife, and the existence of spirit and soul in human beings. As these qualities inform the educational philosophies of both Islam and Christianity, the multi-faith forums intended to produce a faith-based education policy that would be acceptable in multi-cultural societies.

Refining vs. creating disciplines

The movement to Islamise knowledge aims to recast the essentially modern disciplines (such as sociology, anthropology, history, and political science) taught at universities in Muslim societies and, further, to set these disciplines within the context of Islamic worldviews. Proponents thus attempt to bring secular epistemologies and methodologies into line with Islamic teachings. Some advocate for refining the Western epistemic representation of these modern disciplines while others advocate for creating new disciplines based on Islamic epistemology and informed by the socio-cultural, political, and economic, ontological realities of the Muslim world. As a starting point, Al-Fârûqî, who drafted a blueprint for the Islamisation project in 1989, demands that Muslim scholars and intellectuals should master and critically assess both secular and Islamic disciplines, surveying the intellectual problems facing Muslims and restructuring the disciplines in the framework of Islamic epistemology (Al-Fârûqî 1989). Mohamed argues that "the modern discipline as such is not refuted but is Islamized; it is not replaced by an alternative Islamic discipline but is modified to be more in congruence with the Islamic outlook" (Mohamed 1991: 285–286). Nasr (1991) makes a similar argument, emphasising that "Muslims need not claim a new science, but merely look upon it from an Islamic perspective" (Nasr 1991: 399). These admonitions encourage activists to (re)formulate disciplinary outlooks in accordance with an Islamic worldview, rather than produce new fields of knowledge.

The Islamisation of knowledge paradigm, then, deals with methodological rather than epistemological issues. From this reformist educational perspective, the mastery of modern disciplines requires skilled theoretical understanding of Western-informed-secular disciplines, critically engaging their epistemology as well as their methodology of knowledge production and transmission. These methods are employed to refine secular disciplines from Western representation, cultural bias, and worldviews and then integrate their refined contents into "the corpus of the Islamic legacy by eliminating, amending, reinterpreting and adopting its components as the worldview of Islam and its values dictate" (Al-Fârûqî 1989: 18). This reinvigoration process – and its expected outcome – attempt to elbow out the coloniality of education and (re)create new forms of knowledge rooted in Islam.

Ashraf, a Saudi Arabian educationist, reverses this proposition. He suggests that scholars and educationists lay an epistemological foundation for the Islamisation project by first formulating Islamic concepts for all disciplinary specialisations and then proceeding to evaluate Western concepts in light of the newly founded Islamic concepts (Ashraf 1991: 81–101). Both propositions prioritise methodology and neglect epistemology, [the latter being] a key issue in any discourse on de-Westernisation, de-secularisation, and de-colonialisation. However, Al-Turabi, a Sudanese thinker and founder of the Islamist state in the Sudan, maintains that the Islamisation of knowledge cannot be fully achieved until community leaders Islamise other areas of [the] Muslim lifeworld (Al-Turabi 2010). Utilising state power, the Islamists under his leadership launched what they called a "comprehensive call to Islam" in which all aspects of Sudanese lifeworlds were subjected to a thorough Islamisation process. The motive behind this comprehensive Islamisation was to build a foundation for an Islamic state by (re)engineering society and (re)shaping its sociocultural values in accordance with specific Islamic interpretation. In this context, Islamisation was a state-sponsored project between 1989 and 2019, when Islamists ruled the country, and it aimed to create a state that Islamised knowledge, economics, law, and society, etc.

These propositions for epistemological Islamic reconfiguration generate enormous enthusiasm among Muslim intellectuals and educationists trying to (re)shape educational policy and practice as a departure from the colonial education legacy manifested in contemporary educational institutions in Muslim societies (AbuSulayman 1993; Al-Alwani 2005; Al-Attas 1985; Al-Fârûqî 1989, 1992; Al-Turabi 2010; Hamad 2013; Khalil 2006). The reconfiguration discourses also level serious criticism against the Islamisation paradigm accusing it of trying to build an intellectual wall around Muslim societies. They also assert that the Islamisation of knowledge methodology relies on an authoritarian control of knowledge and therefore can only perpetuate, not resolve, the intellectual predicaments facing the Muslim world (Bukhari 2019; Mabud 2016; Nasr 1991; Paya 2022).

Some thinkers completely reject the Islamisation of knowledge as an epistemic paradigm for educational reformation, an instrument for decolonisation, and a methodology of transformation. Such detractors believe that the contents and methodologies of these academic disciplines are too deeply permeated by secularist, materialist, and colonialist epistemic values (see Sardar 2004: 201) to become successfully Islamised. Sardar problematises the very notion of Islamising modern disciplines, such as sociology, anthropology, psychology, political science, arguing that these were born within the matrix of the Western worldview and that their theories and epistemic contents were thus shaped by Western cultures and reasoning. He writes that "the division of knowledge into the various disciplines that we find today is a particular manifestation of how the Western worldview perceives reality and how Western civilization sees its problems" (ibid.: 192). This argument provincialises modern disciplines and highlights their incompatibility with non-Western civilisations.

The Islamisation of knowledge critique refutes the presentation of Western epistemology as a universal paradigm that can be translated into different cultures. Alternatively, the Islamisation trend presents a new model of social science based on Islamic premises. For example, the movement disapproves of the wide recognition and acknowledgement of Auguste Comte (1798 – 1857), whose ideas – as the [supposed] father of social sciences – were fundamental to the development of social sciences in Western epistemology: instead, it sees Ibn Khaldun (1332 – 1406) in this role. Connell criticises such undertakings, arguing that this alteration offers only rhetorical satisfaction as it is futile to challenge Western epistemological "predominance by discovering alternative 'founding fathers' of the same social science" while embracing the discipline's methodology and scientific approach to knowledge production (Connell 2007: xi). However, the Islamisation of knowledge project goes beyond merely identifying an alternative father to social science. It engages broader epistemological questions that deal with the theorisation of Muslim thought, education, and civilisation, and proposes a new methodological approach imbued with [an/the] Islamic worldview to handle these epistemic problems. The critical question here is: to what extent is the Islamisation of knowledge paradigm an appropriate theoretical, methodological tool for addressing Muslim educational dilemmas and the epistemic injustice produced by European colonialism?

The movement's preparation to recast the discipline of anthropology in accordance with Islamic values provides one example of the issues raised during this process. Some Islamisation of knowledge enthusiasts perceive Western anthropology as a product of colonial enterprise. After all, the Western scholars trained in this discipline played a crucial role in aiding colonial administrations in establishing new rules over the colonies in Africa, Asia, and the Americas. For example, Ahmed (1986) does not reject the discipline of anthropology as such, but rather tries to correct its prejudices, eliminate its shortcomings, and redress its methodological bias as well as its racialised concepts of otherness and exotic representations. The ethnocentrist view of reality, which dominates anthropological works in the West, as well as the categorisation of colonised societies in terms of their ethnic characteristics are among Ahmed's key critiques of Western anthropology. He does not trace the origin of anthropology back to European colonialism, but rather to the rich history of Islamic intellectual tradition. He reads the discipline of anthropology comparatively against Islamic historiographical scholarship, arguing that the discipline of anthropology has content which corresponds with features present within the Islamic intellectual legacy. However, anthropology as a disciplinary specialisation did not exist in the Islamic scholarship legacy under the same designation. Ahmed (1986) argues that anthropology is not actually a creation of Western imperialism, as many assume:

> "The work of Ibn Khaldun is reflected – with theoretical frame and supporting data – in that of some of the most influential contemporary Western theorists including Karl Marx, Max Weber, Vilfredo Pareto and Ernest Gellner. Weber's typology of leadership, Pareto's circulation of elites, and Gellner's pendulum swing theory of Muslim society betray the influence of Ibn Khaldun. It is indeed a tragedy that the science of sociology or anthropology did not develop after Ibn Khaldun" (Ahmed 1986: 56).

Islamisation of knowledge enthusiasts categorise other classical works as anthropological, including sociological works by al-Birûnî (973 – 1048), Ibn Battûta (1304 – 1368), and al-Mas'ûdî (896 – 956) because of the methods these scholars applied to the process of data collection and analysis. These classical scholars conducted extended participant observation of cultures other than their own and employed comparative methods of data collection and analysis in their ethnographic works. Thus, Ahmed (1984, 1986) identifies al-Birûnî as a father of anthropology, substantiating his argument with the example of al-Birûnî's study of Hindu as well as the research methodology that he adopted [devised] during that historical scholarship. In contemporary classification terms, Ahmed categorises al-Birûnî's classical work as anthropology par excellence in both epistemological and methodological terms. This position does not deconstruct anthropology as a discipline but (re)directs its methodology of otherness [and othering] toward fairness and equity, urging anthropologists to record reflexively Muslim societies as they are in time and space – and to avoid distortion and misrepresentation. In practice, Islamic universities have different attitudes toward the discipline of anthropology. During fieldwork in the Sudan's three main Islamic universities between 2019 and 2020, I observed that curriculum designers either dismiss the discipline of anthropology as a colonial epistemic instrument for Western domination or incorporate it into the discipline of sociology, anchoring its foundational roots in classical Islamic scholarship.

Unlike Ahmed, who supports disciplinary appropriation, other theorists of pro-Islamic epistemology oppose it. For example, Sardar proposes that Muslim scholars and educationalists "develop their own disciplines based on their own cultural context and geared to solving their own problems" (Sardar 2004: 200) instead of Islamising Western-oriented disciplines, such as anthropology, which was historically born out of colonialism. Islamising disciplines, Sardar continues, is not a viable epistemic project when one considers the conflict between secular-oriented knowledge and the vision of Islam that seeks to produce knowledge grounded in the ontological realities of Muslims. Sardar worries that the appropriation methodology adopted by some Islamisation of knowledge theorists and their one-dimensional prescription for epistemic reformation "is tantamount to a cosmetic epistemological face-lift and nothing more" (ibid.: 201). He substantiates his reluctance toward disciplinary appropriation by arguing that those attempting to recast these disciplines in accordance with Islamic axiology are already infused with secularist ethics, a materialist metaphysical disposition and positivist epistemic orientations. Thus, this (re)appropriation, at best, "would perpetuate the dichotomy of secular and Islamic knowledge that you [Muslims] are so keen to transcend" (ibid.: 201) in contemporary disciplinary taxonomies.

The nomenclature of social science disciplines is problematic in the Islamisation of knowledge discourse because of the differences in disciplinary taxonomies between Islamic and Western epistemology. Al-Alwani argues that many of the thematic subjects studied today in the social sciences were treated in classical Islamic scholarship through the medium of *fiqh* (jurisprudence and Islamic laws) or *fiqh al-Aḥkâm al-Sulṭâniyyah* (the precepts of power). He gives examples of categories such as systems of governance, political theory, international relations, laws, social welfare, and methods of political analysis that are studied in universities as part of contemporary knowledge (Al-Alwani 1990: 9). Here, the epistemological challenge is how to (re)establish a sound disciplinary foundation for social sciences while also taking into consideration the classical legacy of Islamic jurisprudence as well as sources of knowledge and the Western disciplinary taxonomies which are prevalent in today's academic institutions within Muslim societies. Educationists usually approach this challenge relationally, by reading and interpreting the past critically in order to expand epistemic boundaries: they deconstruct modern disciplines theoretically and methodologically so as to (re)construct them in the light of Islamic epistemology. In this disciplinary (re)arrangement where educationists benefit from all forms of knowledge, Islamic epistemology becomes a centre of relationality and a methodological lens through which the (re)arrangement process passes into the new state of Islam-centred disciplines.

However, such (re)arrangement has limitations. It is a daunting challenge to produce reliable academic materials to cover all areas of the thematic subjects studied in modern disciplines. In this regard, Al-Alwani (1990) argues that Muslim scholars and educationists have difficulty in dissociating sociological aspects from theological aspects. Because there is no clear distinction between them in Islamic sources of knowledge, it is impossible to draw "well-defined divisions among these aspects in the same way that there are divisions among the various contemporary social sciences" (Al-Alwani 1990: 10). Taking the example of political science in Islamic universities, Al-Alwani explicates that the departments of Islam-centred political science face a tremendous challenge "in presenting material on political science in the Islamic tradition in a methodical manner befitting the educational and academic purposes for which they were established" (ibid.: 10). Such a predicament creates an intellectual void and further exacerbates the crisis of thought and epistemic dichotomies (secular vs. Islamic) in Muslim academic institutions. Akram argues that the discipline of political science in Islamic epistemology is a branch of theology since it imperatively depends on spirituality and cannot therefore inspire independent intellectual speculation (Akram 2007). In this sense, the distinction between secular and spiritual domains becomes meaningless in Islamic epistemological discourse.

However, the most problematic dimension of this debate is social dynamism, which necessitates changes in epistemologies and methodologies. According to Sardar, the secular-oriented epistemology of social sciences demands that "man should not believe in any predetermined code for a society but accept the principle that society is a continually evolving and changing phenomenon. There is nothing permanent or unchanging about human nature" (Sardar 2004: 196). Our conditions as human beings are transient and should be theoretically and methodologically comprehended via transient concepts. This social dynamism applies to values, moralities, codes of conduct, practices, modes of thinking and belonging, ethics, and many different aspects of human behaviour and life. In this sense, thought, practice, and knowledge are socially constructed and therefore subject to constant de-construction, review, and re-construction. This ever-changing essence of lifeworlds (re)shapes epistemic practices and policies in the domains of both secular and Islamic education. However, Sardar opines that scientific-informed dynamism in values and moralities contradicts that of the Islamic positionality on ethos, arguing that "Islam teaches that moral values do not change; hence Truth, Goodness, Righteousness, Mercy are constant factors" (ibid.: 196). Protagonists of Islamisation negate such positions by citing the popular phrase, "*al-Islâm ṣâliḥ li kuli zamân wa makân*" (Islam is relevant for all times and places), which is attributed to one of the prophetic sayings conveying the notion that Islam is a comprehensive way of life. In practice, enthusiasts of Islamisation attempt to realise this claimed relevance and settle this dynamism debate through analogical extension and juristic speculations such as *ijtihâd* (disciplined and creative intellectual effort) and *qiyâs* (reasoning by analogy). However, this analogical extension comes, epistemologically, more under the purview of theological discourse than within the sociological discussion of Islamic knowledge and its relevance today.

In the Islamisation of knowledge discourse, natural sciences are perceived to be less problematic because they deal with nature, unlike social sciences whose domain of inquiry is culture, which is always subject to interpretation. Thus, little effort is made to (re)appropriate and (re)arrange the natural sciences' [curricula] to suit Islamic epistemology. Islamic universities in the Sudan operationalise the Islamisation process in faculties of natural and life sciences parallel to respective disciplinary specialisations by simply attaching courses on Islamic studies. The academic contents of natural and life sciences' disciplinary courses in Islamic universities remain unaltered and therefore do not differ thematically from those of non-Islamic universities' courses. Islamic courses are added because Islamic universities intend to produce polymath graduates who are not only qualified in their respective fields of study, but also knowledgeable in Islamic studies. However, some lecturers perceive such a polymathic methodological approach as an academic burden on students in the sense that it distracts them from concentrating on the disciplinary specialisations on which they would build their future careers. In this regard, one lecturer from the Faculty of Pharmacy of Omdurman Islamic University stated, "I don't understand why my students must take courses on Islamic studies because they have not enrolled [at] the university to become religious scholars[,] but pharmacists". This burden feels even heavier among students who are required by university administrations to study Islamic subjects regardless of their disciplinary specialisations or personal interests.

Furthermore, life sciences such as biology, whose epistemology drives from the evolutionary paradigm of knowledge production, are seen as more insidious than natural sciences like physics. The Islamisation of knowledge project severely criticises the evolution theory's influence on biology because of its acute opposition to the idea of God's creation of (in)animate [entities] and the universe at large (see Nur 2022). This kind of conflict questions the viability of the Islamisation project and the suitability of its methodological approach in dealing with the epistemological reconfiguration of sciences in general and natural sciences in particular. Critics of the Islamisation project point to the (in)compatibility of natural sciences and Islamic epistemology. For instance, Sardar argues that "natural sciences are conceived as antithetical to nature, which can be twisted and tortured in the name of progress". In contrast to this scientific outlook, Sardar writes "Islam […] does not encourage confrontation between man and nature. It teaches man to be 'natural' and thus work in harmony with nature" (Sardar 2004: 196). The Islamisation project has yet to practically address this kind of difference in epistemic orientation and outlook.

The Islamisation trend regards knowledge of the world as equally important to faith-based knowledge. Proponents urge Muslim scholars and educationists to understand the scientific world completely and integrate it into the main corpus of the nascent Islamic epistemology. Islamisation of knowledge theory is based upon the unity of truth, which devolves from the oneness of God. Its epistemic orientation rests on the idea that the truths conveyed in divine revelation cannot be different from those of reality. This divine truthfulness claim implies that if a researcher has conducted scientific research and yielded findings that contradict any element of the divine reality, then those findings must be reworked or considered false. In this sense, Islam is defined in religious terminologies and employed as a yardstick against which to measure epistemic practices on the ground and their (in)compatibility with the dictum of the nascent knowledge politics.

Islamisation of knowledge debates: Concluding remarks

The Islamisation discourse was born out of a reaction to colonial, cultural and epistemic invasions, the secularisation of academic institutions, and Orientalism (Abaza 2000) and its misrepresentation of Islamic knowledge as anachronistic and unfit for scientific modernity. Other factors that triggered the Islamisation of knowledge debate are the decline of Islamic epistemology and Muslims' current uncritical attitude of mimicry in knowledge practices and modes of thinking rooted in other cultures and civilisations, especially

Eurocentrism with little or no critical engagement. Western-oriented education introduced to Muslim societies during European colonialism seems to have played an important role in creating this sense of mimicry, alienation, and estrangement. Moten (1990) also highlights the role of nostalgia in this process. According to Moten, Islamic epistemology, which once imparted life and motion to a great leading civilisation, is today confined to antiquated schools of thought and classical books (Moten 1990). As a critical response to intellectual modernity, the Islamisation discourse raises legitimate questions about epistemic emancipation, the decolonisation of knowledge and mind, and the indigenisation of social and natural sciences. Critical engagement in these issues was not only a concern of Muslims and their reaction to Orientalist scholarship, but also the subject of apprehension in many societies in the Global South who underwent similar arduous colonial experiences (Falola 2022; Freire 1998; Mudimbe 1988; Ndlovu-Gatsheni 2018; Ngũgĩ wa Thiong'o 1981; Smith 2012; Santos 2014, 2018). Thus, the knowledge produced in the matrix of Islamisation served as an instrument for liberation, epistemic justice, restoration of authenticity, self-representation, comprehensive and self-contained philosophy for life. It is a form of knowledge born from the struggle against historical colonialism, the current state of coloniality, and its epistemic genocidal policy and hegemonic power apparatus.

The Islamisation of knowledge project has imbued the Muslim psyche with a sense of epistemic mission whose actualisation became increasingly urgent in many societies that prioritised a spiritual approach to scientific inquiry. Insurgent decolonial scholarship sees this faith-based approach to knowledge as an instrument for divesting modern disciplinary fields in academia and methodology of their Western representations and recasting them with an Islam-centred epistemology. Nasr argues that the Islamisation project "has produced a plethora of academic and ideological expositions on the place of Islam in intellectual inquiry and has ascertained the claim of Muslims to draw the boundaries of human knowledge anew" (1991: 387). The plethora of intellectual production in this nascent epistemological undertaking has "extended the purview of the Islamisation of knowledge project into every nook and cranny of academic thought" that encompasses social, natural, and life sciences (ibid.: 387). However, the epistemological claim of producing new forms of knowledge grounded in Islamic worldviews remains a tenuous proposition in Islamisation narratology because most of the decolonial claims in this discourse remain undelivered or unconsolidated into solid scientific gains. As a result, the Islamisation project is more of a theoretical intellectual exercise than a practical solution. How Islamisation is actually performed and how it contributes to decolonisation remain unrealised. Adding the adjective "Islamic" to a discipline or adding Islamic courses to a disciplinary specialisation does not automatically transform the discipline's epistemic orientation, as protagonists of the Islamisation of knowledge might think. The academic contents must undergo a thorough systematic review and methodological reconfiguration and (re)establishment to be considered Islamic and to reflect the essence of Islam-centred epistemology.

One of the foundational critiques of Western-centred epistemology on which the Islamisation of knowledge project builds its epistemic urgency and reform arguments is Comtean sociological positivism, which offers the idea that social phenomena can operate and be investigated in a systematic manner similar to that of physical phenomena. This positivist paradigm emerged historically in the matrix of the Enlightenment thinking in the 19th century when Europeans began to question despotism in religious institutions and governments. New Enlightenment ideas led to massive scientific advancement in the natural sciences. The Enlightenment's foundation in methodological empiricism inspired social scientists to adopt similar scientific methods in their sociological investigations of societal problems. Under the influence of the positivist paradigm of scientific empiricism, theorists of social sciences and humanities tried to exclude non-observatory and non-sensorial ways of knowing such as faith-based epistemology, theology and intuition seeing these as vacuous kinds of knowledge. The Islamisation of knowledge trend regards the epistemic position that separates reason from revelation as reductionist, maintaining that non-observatory forms of knowledge, feelings, values, desires, beliefs, and transcendental thoughts are essential epistemological components.

While positivists reject metaphysical components in favour of sensory experiences or reasoning, ultimately, there are no such things as value-neutral knowledge and universal objective truth. Historical and social conditions inevitably affect the scholarly representation of social ideas. Moten argues that value-neutrality in knowledge production is a myth because values, of whatever kind, provide a matrix that (re)shapes the selection of thematic subjects for investigation, formulation of theoretical concepts, and selection of data for analysis and interpretation (Moten 1990: 169–170). The nascent Islamisation project adapts this kind of reflexive approach to refute the claims of positivists and critical rationalists. Proponents provincialise Western epistemology, pointing to the many interpretations of social phenomena that derive from various epistemological understandings and intellectual traditions. The resurgent Islamisation paradigm attempts to put Muslim perspectives on knowledge production at the centre of epistemic orientation without necessarily excluding other intellectual traditions in its knowledge practices. Ngũgĩ wa Thiong'o argues that the basis of absorbing the world is to know oneself and one's environment. This mode of thinking is part of a common desire in the Global South to move "the center from its assumed location in the West to a multiplicity of spheres in all the cultures of the world" (Ngũgĩ wa Thiong'o 1993: xvi). This insurgence of the marginalised assigns epistemic legitimacy and centrality to all cultures, thus emphasising plurality, multiplicity, and relationality as forces of dynamism and exchange.

The epistemological perspective of positivism led gradually to the rise of secularism in Europe, influencing the methodology of social sciences and research orientation. This paradigm shift instigated the development of secular-scientific ideology in the West and (re)shaped the politics of sociological knowledge production and ways of thinking not only in Western societies, but also in other societies that had fallen under Western hegemony during European colonialism in the 19th and 20th centuries. This encounter resulted in an intellectual invasion of colonised societies and the marginalisation of their knowledge systems. While the positivist paradigm attempts to build an exclusive methodology based upon human reason, disregarding religious thoughts and the sacral legitimation of political power and authority, the Islamisation paradigm emphasises Islam-centred epistemology, examining social phenomena and individual behaviours within the context of a broader social system and considering the observable and non-observable dimensions of human experiences (Moten 1990). The reliability and comprehensiveness of this Islamic paradigm in addressing methodological problems and building an epistemology informed by Islamic worldviews and ontologies depends on how it conceives Islam and operationalises it in the process of knowledge production. This is because any attempt to (re)shape epistemology in accordance with the Islamic legacy raises scepticism about the nature of this legacy as well as the conflicting constitutive elements upon which theories of knowledge are constructed. As Mohamed (1991) argues, the Islamic legacy contains variant epistemological perspectives that emerge from conflicting schools of thought and different socio-cultural, political, and historical contexts (Mohamed 1991: 290). It is not a monolithic system of thought, as some protagonists of Islamisation of knowledge might suggest, but rather a multiple and relational body of ideas that allows for different approaches to knowledge.

The Qur'an, Sunnah or *ḥadîth*, *ijma'* (the juristic consensus), and *qiyâs* (reasoning by analogy) are the main sources of Islamic knowledge in the Islamisation of knowledge project in the Sudan, in Africa, and across the Muslim world. In the debate over the Islamisation of knowledge, two main factors weaken a potential epistemology based [solely] on these four sources of Islamic knowledge. First, this debate implicitly defines Islam as a religion and therefore imposes an epistemic reductionism and limitation constraining the enhancement of knowledge. Second, subjective readings of the Qur'an and Sunnah as well as other sources yield multiple interpretations. These varied interpretations would inevitably produce variant worldviews and epistemic orientations and thus foster heterogeneity in religious thoughts. However, heterogeneity does not equate to a comprehensiveness in terms of ideas because the broader framework within which this multiplicity manifests itself is a religious edifice. Moreover, some scholars express scepticism toward the connection between knowledge and religion, arguing that it could create a metaphysical science that would impose religious oppression and systematic censorship on scientific ideas and scholarship in the name of Islamic authenticity (see for example, Abaza 2000). Such concern is legitimate, especially when protagonists of Islamisation see Islam as a religious system of thoughts with a clear epistemological delineation and methodological measures of knowledge assessment and validation. Such a conception is reductionist in the sense that it tries to trace the roots of all knowledge back to religion. The danger of this notion is that it has the potential to create a serious methodological shortcoming that would further eclipse Islamic epistemology. Moten argues that the soundness of Islam-centred epistemology rests on an intellectual openness that perceives Islam as a civilisation containing philosophies, religious thoughts, various forms of knowledge, and intellectual traditions. This totality "looks upon life as an organic whole and approaches its problems in the light" of societal values and ontological realities (Moten 1990: 163) and has an ability to draw from multiple sources and different, perhaps even contradictory, forms of knowledge in its construction of nascent epistemology.

Nasr argues that the Islamisation of knowledge project "has been shaped more in the spirit of political discourse" than by a constructive and reflexive academic dialogue (Nasr 1991: 387). This statement causes the Islamisation of knowledge policy and practice to resonate in a country like the Sudan where Islamisation was a state-sponsored project during the Islamist reign from 1989 until 2019. This is also applicable to a country like Malaysia, which is witnessing educational reconfiguration and the (re)construction of a new state discourse on science and Islam. In Malaysia, protagonists Islamise state institutions, including academic ones, and enhance new bureaucratic elites who share a common Islam-informed ideology. Abaza states that the Islamisation of knowledge debate in Malaysia is at the centre of political power because protagonists of this epistemic paradigm occupy influential positions in academic institutions, publishing, and governmental offices (Abaza 2000: 62). In secular-oriented countries such as Egypt, however, the promotion of the Islamisation of knowledge occurs via political activism in which Islamic groups, organisations, and individuals play a dynamic role in negotiating the nascent epistemology project through society. Thus, the policy and practice of the Islamisation of knowledge on the ground differ from one society to another depending on socio-political contexts and prevailing educational circumstances. However, asymmetry in power relations between the West and the Global South and its hegemonic impact on educational policy are significant factors in the instigation of the Islamisation of knowledge discourse.

Acknowledgements

This article draws from a field research in the Sudan in 2019 and 2020 funded by the Fritz Thyssen Foundation to which I am grateful. Writing up the article and the development of its arguments took place during a postdoctoral fellowship project financed by the Africa Multiple Cluster of Excellence at the University of Bayreuth (funded by the German Research Foundation under Germany's Excellence Strategy – EXC 2052/1 – 390713894). I presented an earlier draft of this article at the Africa-Bayreuth International Alumni Network (afriBIAN) conference "The Politics of Religion in Africa," held on November 1 – 4, 2022, at the University of Jos in Nigeria. I am thankful to the organisers of these conferences for supporting my work and for the constructive discussions in the regular meetings as well as the fruitful informal conversations during the breaks. Thanks go also to the editors of this volume for their invaluable comments on the article.

Bibliography

Abaza, M. (2000): "The Islamization of Knowledge between Particularism and Globalization: Malaysia and Egypt." In: C. Nelson/S. Rouse (eds.), Situating Globalization: Views from Egypt, Bielefeld: transcript Verlag, pp. 53–95. https://doi.org/10.14361/9783839400616-003.

AbuSulayman, A. (1993): Crisis in the Muslim Mind, Herndon: International Institute of Islamic Thought.

Ahmed, A. S. (1984): "Al-Beruni: The First Anthropologist." In: Royal Anthropological Institute of Great Britain and Ireland, 6, pp. 9–10. https://doi.org/10.2307/3033407.

Ahmed, A. S. (1986): Toward Islamic Anthropology, Herndon: International Institute of Islamic Thought. https://doi.org/10.35632/ajis.v3i2.2893.

Akram, E. (2007): "Muslim Ummah and its Link with Transnational Muslim Politics." Islamic Studies 3, S. 381–415.

Al-Alwani, T. J. (1990): "Political Science in the Legacy of Classical Islamic Literature." The American Journal of Islamic Social Sciences 1, pp. 9–14. https://doi.org/10.35632/ajis.v7i1.2663.

Al-Alwani, T. J. (2005): Issues in Contemporary Islamic Thoughts, Herndon: The International Institute of Islamic Thought. https://doi.org/10.2307/j.ctvk8w1ww.

Al-Attas, S. M. (1985): Islam, Secularism and the Philosophy of the Future, London/New York: Mansell Publishing Limited.

Al-Fârûqî, I. R. (1989): Islamization of Knowledge: General Principles and Work Plan, Herndon: International Institute of Islamic Thought.

Al-Fârûqî, I. R. (1992): Al Tawḥîd: Its Implications for Thought and Life, Herndon: International Institute of Islamic Thought.

Al-Turabi, H. (2010): Fîl Fiqhî as-Siyâsî: Muqârabât fî Ta'ṣîl al-Fikr as-Siyâsî al-Islâmî (Approaches to the Authentization of Islamic Political Thought), Beirut: Arab Scientific Publishers, Inc.

Ashraf, S. A. (1990): Islamic Education Movement: An Historical Analysis, Cambridge: The Islamic Academy.

Ashraf, S. A. (1991): "Islamic Education and Evaluation of Post Conferences and Post Conference Achievments." In: F. Malkawi/H. Abdul-Fattah, The Education Conference Papers, Vol. 1, Amman: Islamic Studies and Research Association, pp. 35–36.

Bukhari, S. M. (2019): "The Islamization of 'ilm: Ideals and Realities in a Globalized World. " In: S. Akkach, 'ilm: Science, Religion and Art in Islam, Ade-

laide: University of Adelaide Press, pp. 98–112. https://doi.org/10.20851/ilm-1-06
Connell, R. (2007): Southern Theory: The Global Dynamics of Knowledge in Social Science, Cambridge: Polity Press. https://doi.org/10.22459/AHR.44.2008.04.
Falola, T. (2022): Decolonizing African Knowledges: Autoethnography and African Epistemologies, Cambridge: Cambridge University Press. https://doi.org/10.1017/9781009049634.
Freire, P. (1998): Pedagogy of Freedom: Ethics, Democracy, and Civic Courage, Lanham & Boulder: Rowman & Littlefield Publishers, INC.
Hamad, M. A.-G. (2013): Manhajiyyat al-Qur'an Al-Ma'rafiyya (Epestimic Methodology of the Qur'an), Beirut: Mominoun Without Borders.
Khalil, '. A.-D. (2006): Maqaddima ila Islamiyya al-M'arifa (Introduction to Islamic Epistemology), Beirut: Dar Ibn Kathîr.
Lo, M./Haron, M. (2016): "Introduction: Africa's Muslim Institutions of Higher Learning: Moving Forward." In: M. Lo/M. Haron (eds.), Muslim Institutions of Higher Education in Postcolonial Africa, Hampshire & New York: Palgrave Macmillan, pp. 1–9. https://doi.org/10.1057/9781137552310_1.
Mabud, S. A. (2016): "World Conference on Muslim Education: Shaping the Agenda of Muslim Education in the Future." In: N. A. Zaman (ed.), Philosophies of Islamic Education, Historical Perspectives and Emerging Discourses, New York & London: Routledge, pp. 129–143.
Manger, L. (2013): "Muslim Diversity: Local Islam in Global Contexts." In: L. Manger (ed.), Muslim Diversity: Local Islam in Global Contexts, London & New York: Routledge, pp. 1–36. https://doi.org/10.4324/9781315027593.
Manger, L. O. (2002): "Religion, Identities, and Politics: Defining Muslim Discourses in the Nuba Mountains of the Sudan." In: Journal of Arabic and Islamic Studies 4, pp. 132–152. https://doi.org/10.5617/jais.4569.
Mohamed, Y. (1991): "Islamization of Knowledge: A Critique." In: The American Journal of Islamic Social Sciences 11/2, pp. 282–294. https://doi.org/10.35632/ajis.v11i2.2439.
Moten, A. R. (1990): "Islamization of Knowledge: Methodology of Research in Political Science." In: The American Journal of Islamic Social Sciences 7/2, pp. 161–175. https://doi.org/10.35632/ajis.v7i2.2788.
Mudimbe, V. Y. (1988): The Invention of Africa: Gnosis, Philosophy, and the Order of Knowledge, Bloomington & London: Indiana University Press.
Nasr, S. V. (1991): "Islamization of Knowledge: A Critical Overview." In: Islamic Studies 30/3, pp. 387–400.

Ndlovu-Gatsheni, S. J. (2018): Epistemic Freedom in Africa: Deprovincialization and Decolonization, New York & London: Routledge. https://doi.org/10.4324/9780429492204.

Ngũgĩ wa Thiong'o (1993): Moving the Centre: The Struggle for Cultural Freedoms, Oxford & Nairobi & Portsmouth: James Currey & EAEP & Heinemann.

Ngũgĩ wa Thiong'o (1981): Decolonising the Mind: The Politics of Language in African Literature, London: James Currey.

Nur, M. B. (2022): "Politics of Epistemology in Postcolonial Africa: The Islamisation of Knowledge in the Sudan." In: Politics, Religion & Ideology 23/4, pp. 475–496. https://doi.org/10.1080/21567689.2022.2139688.

Paya, A. (2022): "'Islamization of Knowledge': A Utopian Dream or a Viable Project? A Critique of the Views of Two Prominent Iranian Proponents of the 'Islamization Project'." In: British Society for Middle Eastern Studies, pp. 1–21. https://doi.org/10.1080/13530194.2022.2048462.

Santos, B. D. (2014): Epistemologies of the South: Justice Against Epistemicide, London & New York: Routledge.

Santos, B. S. (2018): The End of the Cognitive Empire: The Coming of Age of Epistemologies of the South, Durham & London: Duke University Press. h ttps://doi.org/10.1215/9781478002000.

Sardar, Z. (2004): Desperately Seeking Paradise: Journeys of a Sceptical Muslim, London: Granta Books.

Shani, G. (2008): "Toward a Post-Western IR: The Ummah, Khalsa Panth, and Critical International Relations Theory." In: International Studies Review, 4, pp. 722–734. https://doi.org/10.1111/j.1468-2486.2008.00828.x.

Smith, L. T. (2012): Decolonizing Methodologies: Research and Indigenous Peoples, London & New York: Zed Books.

Towards Parity in Knowledge Production within the Framework of North-South Collaboration

Michaela Pelican and Jonathan Ngeh

Introduction

A critical aspect in the struggle for recognition and inclusion of colonised and formerly colonised peoples – including indigenous peoples and ethnic/racialised minorities – relates to questions of representation. In particular, we pay attention to the question of how they are represented or silenced in historical accounts and excluded from knowledge production and dissemination, even when such knowledge directly relates to them. In view of this debate, the hegemony of Euro-North American-centric thought and viewpoints has been subjected to extensive critique from anti-, post- and decolonial theorists for its reinforcement of neo-colonial relationships and structural inequalities (Bhabha 2004; Fanon 1990; Mbembe 2016; Ngũgĩ wa Thiong'o 1986; Said 1979; Smith 1999).

We see the debate on the decolonisation of higher education today as a reflection of the historical struggle for recognition and inclusion of colonised and formerly colonised peoples. Furthermore, we view knowledge production as being deeply implicated in power struggles and agree with Linda Smith's assertion that research is a significant site of struggle between competing interests (Smith 1999).

Academic research enjoys a near monopoly in knowledge production, which is why the way that we conduct research is also the focus of the analysis offered here. Universities and funding organisations play a central role in knowledge production by providing the resources, infrastructure and manpower among other key requirements. Consequently, the critique of Euro-North American centric epistemology and its hegemony applies to the established process of research and the institutions that reproduce it. Therefore, it is important to consider both the process and the institutional context in analyses of knowledge production.

Our aim is to explore a collaborative approach to knowledge production and dissemination that fosters inclusion and parity by embracing a diversity of experiences and perspectives in the research process. Furthermore, we aim to interrogate existing hierarchies and subjectivities that may perpetuate the marginalisation and othering of historically marginalised peoples. Specifically, we focus on collaboration within a research team comprising academic staff at various stages of their careers and students, both undergraduates and graduates, from four universities, three of which are in Cameroon (Bamenda, Dschang, Yaoundé 1) and the fourth in Germany (Cologne). We view the project partners based in Cameroon as occupying the dual position of outsiders and insiders in the research process since they are both researchers and members of a formerly colonised society (Cameroon). By including Cameroonian and German researchers on par, we allow for perspectives grounded in the South and the North to shape the research process and outcome. However, decolonial critics call for recognising knowledge production that occurs outside of academia: in our case this would have required the inclusion of members of the society under study beyond the collaboration partners in Cameroon. While we agree with this argument, it is beyond the scope of the project we discuss here.

Our research is guided by the notion that knowledge production is a "collaborative process" involving various actors who are positioned differently. In order to minimise asymmetrical power relations and the exclusion that can arise from it, we draw on a "collaborative process": "researching with and for people rather than on people" (Mitlin and Thompson 1995: 238). This approach, known as the participatory methodological approach, calls for collaboration, recognition and inclusion in the entire research process. We explore this approach in a research project that brought together colleagues and students from universities in Cameroon and Germany in the summer of 2018 to work together in Yaoundé, Cameroon. Conceptually, we draw on Homi K. Bhabha's (2004) concepts of hybridity and Third Space. The chapter is organised in the following manner: The introduction is followed by a discussion on research collaboration in the asymmetrical North-South partnership, and the outline of our conceptual framework. The ensuing sections are devoted to empirical analysis and concluding remarks.

Power asymmetry and partnership in knowledge production

A critique of the hegemony of Euro-North American centric epistemology from decolonial and postcolonial perspectives is that it is rooted in an epistemology that excludes the very people who are the target of scientific inquiry: "the exclusion of the people talked about, from the discussion about them" (Hountondji 2005: 531). Additional criticism arising from a post-structuralist perspective holds that the hegemony of a Euro-North American centric epistemology is rooted in asymmetrical power relations that silence or dismiss the worldviews of colonised and formerly colonised peoples as well as other minorities (Hall 1992).

A common thread in the decolonising and postcolonial approaches to tackling the above problem is the call to place colonised and formerly colonised peoples and their own experiences and thoughts at the centre of knowledge production. Thus, from the position of Maori indigenous people, as explained by Smith (1999), decolonisation is initially about centring the concerns and world views of the Maori and then coming to know and understand theory and research from the Maori perspective and utilising these for Maori purposes. Likewise, drawing from African experience, decolonisation is about a liberating perspective that allows Africans to understand themselves clearly in relation both to one another and to other people in the world (Ngũgĩ wa Thiong'o 1986). Central to the arguments provided by these authors is the desire for formerly colonised and indigenous people to champion the research about their community and people (de Sousa Santos 2007; Ndlovu-Gatsheni 2018; Nyamnjoh 2012). That said, proponents of the decolonisation perspective warn that decolonisation is not the complete rejection of Western knowledge, but rather it is (among other things) about the critical examination of the Western hegemonic methodologies and perspectives, and evaluation of their relevance before any application (Hountondji 2005; Ngũgĩ wa Thiong'o 1986).

The call for colonised and formerly colonised peoples to champion research about themselves and their communities, and to place their experiences and worldviews at the centre of knowledge production is undoubtedly crucial in any attempt to counter the Western hegemony of knowledge. In practice, we believe that this goal cannot be easily achieved by minority scholars or scholars from the South alone precisely because knowledge production is intertwined with global economic and political structures of inequalities that perpetuate colonial power relations and intersect with local power asymmetries. Moreover, knowledge production is costly and research in the Global South is severely underfunded and over-reliant on external donors in the Global North who impose their own standards of accountability, responsibility and measures of success. The implication of this, as indicated in the situation of Cameroon (Ngeh 2021; Walter et al. 2017) and other African countries (UNESCO Institute for Statistics 2015), is that donors use their leverage to shape the nature and focus of research on the African continent.

In anthropology, the problem of excluding the researched from the key processes of the research on them was first addressed through the adoption of reflexivity during anthropological research undertaken during the 1940s. Reflexivity entails a form of ethnographic writing that goes beyond simply reporting information provided by the subject of inquiry and instead includes the social discourse of the people under study (Wolf 2000). It allows for the incorporation of the perspective of those under study into the knowledge produced about them. In the 1970/80s, anthropology experienced a new reflexive turn which questioned the objectivity of anthropological research and called for the critical interrogation of the researcher's subjectivity and positionality in relation to the people under study. While the reflexive turn did not go unchallenged, it has opened the way for a more critical and engaged anthropology that pays attention to intersubjectivity and the co-production of knowledge (e.g., Davies 1998; Scholte 1974). In later years reflexivity was criticised for insufficiently addressing the power dynamics that privilege the dominant epistemology and critics called for a collaborative and symmetrical approach that would allow the researched to be involved as much as possible throughout the entire research process (Aijazi et al. 2021; Gay Y Blasco and De La Cruz Hernández 2012; Martin and Dandekar 2022). Notwithstanding, Aijazi and colleagues (2021) note that while the motivation for instituting collaborative research arrangements might be driven by concerns for parity, it can be sometimes "rooted in a calculus intended to render ethnographic research more convenient and conducive to the changing expectations and roles of scholars in Western universities" (ibid.: 75). In this regard, collaborative research primarily serves the interests of the university and faculty hosting the research project, and not necessarily research partners or research communities. It reinforces existing hierarchies and unequal power sharing within research teams. Furthermore, we realise that the discourse of research collaboration is riddled with contradiction. The call for collaboration in academia is at odds with the systematic devaluation of the outputs from such research. In the social sciences and humanities, co-authored publications, as noted by Aijazi et al. (2021), are often ranked below single-authored articles for merit, tenure and promotion purposes. We view this contradiction as an example of institutionalised practices that normalise established hierarchies and neo-colonial practices of exclusion in research.

The work of Blasco and Hernández (2012) stands out as an excellent example of symmetrical collaborative partnership. It is a co-authored publication by a researcher and research subject, which includes both perspectives on the research process. We draw on the work of Aijazi et al. (2021) on collaboration which recognises power asymmetry and tension in the research partnership as unavoidable problems that must be confronted. Following Blasco and Hernández (2012) and Aijazi et al. (2021) we argue for a collaborative and inclusive approach that interrogates power asymmetry; contests the practices that maintain dominant cultural ideologies; challenges the North/South, Us/Them, coloniser/colonised etc. dichotomies; challenges the hegemony of the worldview of the dominant group. Our approach to collaborative research is further informed by the notion of hybridity and Third Space (Bhabha 2004) in postcolonial studies, topics which we outline in the next section.

Homi K. Bhabha's Third Space and hybridity

"Hybridity" refers to the mixing of cultures, while the concept of Third Space concerns itself with addressing the space, developed between two poles or binarities — self /other, colonisers/colonised etc. (Bhabha 2004). Bhabha explains that the notion of hybridity derives from the idea of translation, understood as "a double process of decontextualization and recontextualization, first reaching out to appropriate something alien and then domesticating it" (Burke and Hsia 2007: 10). Cultural translation therefore opens up the possibility for something new, while in the same process it "denies the essentialism of a prior given original or originary culture" (Bhabha 1990: 211). This optimistic assessment of hybridity has been linked to the Latin American concept of *mestizaje* which, like hybridity, celebrates processes of mixture (Wade 2004). The concepts of Third Space, hybridity and *mestizaje* have been celebrated as the antidote to essentialist ideas of "race" and culture (Fernandez 1992; Ghasemi et al. 2018).

However, both the concepts of hybridity and *mestizaje* have been criticised for assuming the existence of hitherto undifferentiated cultures or knowledges, which is the problem that the authors of these concepts initially sought to counter (Howell1996; Wade 2004). Bhabha has been criticised by Zhou and

Pilcher (2019) for assuming that the Third Space consisted of symmetrical relations. Zhou and Pilcher (2019) note that what seems to elude the Third Space discourse is a discussion of what this space really is, and what the meaning of the word "third" actually entails. Instead, Zhou and Pilcher (2019) conceptualise the Third Space as a "moment of intervention" that addresses power structures (ibid.: 5). This argument is echoed by Wolf (2000) who also considers reflexive ethnography to be relevant to practices of intervention in the Third Space. A reflexive ethnography, as he explains, calls for an ethnographic writing that is open to a plurality of voices that allows for a collective construction of knowledge.

In light of the discussion above, our analysis draws on the Third Space as a moment of intervention that is critical, participatory, and emancipatory. This allows for an intervention in the process of knowledge production at both individual and intercultural levels. The individual level refers to a moment of critical reflection that allows us to interrogate assumptions about cultural differences of self and others. And the intercultural level builds on an intervention strategy that involves interrogating hierarchical structures and dominant discourses that "otherwise" and silence postcolonial subjects, ethnic minorities and the oppressed (Aijazi et al. 2021; Martin and Dandekar 2022; Zhou and Pilcher 2019). The challenge here, as noted by these authors, is that power struggle remains an inescapable and often uncomfortable facet of individual experiences within intercultural dialogue. The opening/closure of intercultural dialogue is intricately linked to the extent to which those involved remain willing to "descend" into the instabilities typical of the Third Space. Furthermore, intercultural dialogue can also occur without the participants crossing boundaries that produce otherness, leaving them in the essentialist and polar end of the Third Space. On the other hand, those engaged in intercultural dialogue can transcend and dissolve essentialist boundaries, bringing them to the non-essentialist version of Third Space (Holliday and Amadasi 2019; Zhou and Pilcher 2019). In terms of method, our analysis draws on critical self-reflection to investigate different experiences in a diverse research team, allowing us to reflect on our own experiences in the project (Khosravi 2007), and to demonstrate the effectiveness of parity in research in the context of North-South collaboration.

Collaboration in knowledge production in a North-South research partnership

The project "Urban youths' perspectives on making a future in Cameroon and/or abroad" was realised in the summer of 2018. For six weeks, twelve students and five supervisors from Cameroon and Cologne conducted fieldwork in the capital city of Yaoundé on the subject of future making under conditions of uncertainty. The students and supervisors formed German-Cameroonian research tandems and tackled the subject from different angles.[1] The projects focused on the strategies of young artists and journalists; the experiences of female university graduates in the labour market; the gendered trajectories of educated women and men of the Mbororo ethnic minority; the challenges faced by un/successful return migrants; and the contributions to youth development by migrant investors and home town associations. The results of our joint research will be published in a forthcoming thematic issue on "Urban Youth and Future Making in Cameroon".

The project grew out of an initial partnership in 2007 between Michaela Pelican and colleagues of the Department of Anthropology of the University of Yaoundé 1, as well as long familiarity and partnership between Jonathan Ngeh and Michaela Pelican dating back to the early 2000s when Jonathan was still an undergraduate student in Sociology and Anthropology at University of Buea and Michaela a PhD student conducting her fieldwork in Cameroon. In 2017, Deli Teri Tize (University of Yaoundé 1) and Michaela Pelican (University of Cologne) developed the idea of a collaborative research project involving students from the two partner universities to conduct research in two sites: the capital Yaoundé in the francophone Centre region and the city of Bamenda in the Anglophone North West. However, due to the political crisis in the Anglophone North West and South West regions, which started in 2016 and took a violent turn in October 2017 (Pelican 2022), conducting fieldwork in Bamenda was no longer feasible. In response, Deli and Michaela decided to limit the

1 The institutions involved were the Universities of Bamenda (UoB), Cologne (UoC), Dschang (UoD) and Yaoundé 1 (UoY1). The students who participated were Nzouenkeu Guylaine and Chancelyne Wulseh Yein (UoB), Dana Harms, Johanna Merz, Eugene Tingwey, Lotta Schütt, Brice Stapelfeldt and Anna Woelki (UoC), Isa Adamu and Hamza Dabo (UoD), and Awah Kum Jr., and Wendon Gillian Mbuh (UoY1). The supervisors were Deli Teri Tize and Afu Isaiah (UoY1), Michaela Pelican (UoC), Jonathan Ngeh (UoB), Alawadi Zelao (UoD).

fieldwork to Yaoundé and to open up the project to students and supervisors from other parts of Cameroon in order to accommodate different perspectives. In a back-and-forth process, they invited colleagues from the Universities of Bamenda (Jonathan Ngeh), Dschang (Alawadi Zelao) and Yaoundé 1 (Afu Isaiah) to jointly refine the project design and bring in students from their respective institutions. The project management rested with Michaela and Deli.

The research theme of future making was further developed on the basis of the shared interests of the scholars involved in the study. While the project aimed at capturing some of the policy changes in Cameroon in the past decade so as to understand their impact on Cameroonian youths and their perspectives of the future, the study also tried to understand current role models and notions of success among urban youths in Cameroon and the effects these factors had on return migrants and migrant remittances to the country. Importantly, the project aimed at promoting collaboration and shared knowledge production amongst students and supervisors based in the Global South and North.

The collaboration project consisted of three phases: The first phase happened remotely and comprised preparatory seminars at the partner universities which were adapted to the respective curricular requirements. While the students at the University of Cologne had the opportunity to attend two seminars geared towards the regional, thematic and methodological preparation of the planned collaborative research which were part of the curriculum, the Cameroonian partner universities were unable to accommodate the research preparation in their curricular structures. The supervisors and students had to make time besides their regular coursework to discuss selected readings and develop research ideas. To build a joint basis for the research collaboration, students in Cameroon and Germany were asked to write summaries of selected readings and share them in the group. They were also required to develop their research ideas in the form of a proposal and share it with all project members. Furthermore, students were encouraged to read all the proposals, identify possible research partners and explore possibilities for working together through one-on-one exchanges via email and WhatsApp. Importantly, German students were encouraged to seek research partnerships with their Cameroonian peers and vice versa.

In the second phase of the project, the students and supervisors teamed up in Yaoundé for a period of six weeks to conduct fieldwork, analyse data from the field, and share the preliminary findings with the local population and key stakeholders. This phase kick-started with an intensive five-day seminar that enabled Cameroonian and German students and supervisors to come to know each other and further develop their joint research interests; to engage with the project's thematic and theoretical framework and advance discussions on an intercultural level to elaborate on their research methodology and finalise organisational arrangements for practical fieldwork. The seminar was followed by a four-week period dedicated to fieldwork. During this phase, the students worked in German-Cameroonian research partnerships. They had the opportunity to meet with their supervisors on a regular basis and sometimes be accompanied to the field. After the fieldwork phase was completed, the data collected was analysed in a five-day workshop. Students and supervisors discussed and synthesised their research findings and prepared a joint research report. The preliminary findings of the research project were presented at a public conference organised at the University of Yaoundé 1 and disseminated on national TV through the participation of selected students in a popular TV show.[2]

Importantly, the project enjoyed the support of the Faculty of Arts, Letters and Social Sciences (FALSH) of the University of Yaoundé 1 which acted as the host university. They facilitated the German partners' visa procedures, welcomed project members and continuously provided infrastructural support, such as halls for seminars and the public conference. Funding for the research collaboration was provided by the Department of Social and Cultural Anthropology of the University of Cologne. The funding covered all research expenses of the Cameroonian and German partners, including transport, accommodation, and allowances for fieldwork. All students received the same fieldwork allowances, a decision that aimed at ensuring that they had equal financial resources. The supervisors received a moderate honorarium for their extra work. The responsibility for the project's accounting rested with the German partner.

2 Thanks to one of our student's personal connections to the Cameroon radio and television station (CRTV), we were invited to present the topic and preliminary findings of our collaborative research project in the morning talk show *Hello Cameroon*, which is broadcast live. When interviewed by the talk show moderator Gwendoline Egbe, Lotta Schütt (University of Cologne) and Hamza Dabo (University of Dschang) talked about Cameroonian youths' perspectives on making a future (*Hello Cameroon*, 30th August, 2018).

Our different positionalities in the project

As is often typical in academic research projects, ours went through distinct phases, starting with the conception of the project and leading through its execution to the completion process. While Michaela and Deli initiated the collaboration and were involved at the earliest stage, the other project members joined the team at different points in the process. The presence of some project members at the start of the project through to its completion placed them in a position to exert more influence on the research and subsequent outcome. Similarly, the University of Cologne and the University of Yaoundé 1 had more influence than the other partner universities by virtue of oversight responsibilities which they enjoyed as funders and hosts of the project. Finally, the relationship between students and academic staff was also hierarchical because of the very different roles that universities in general assign to students and academics. For example, the reports of students were evaluated and graded by the supervisors, an indication of the influence they have over students.

In terms of perceived power and privilege, our social positions as academics or students differed from each other because of our ascribed racial background (European and African), citizenship, gender, sexual orientation, age, and achieved status (professor, associate professor, lecturer, instructor and student). The circulation of power within the project took different forms, with some overt and some less obvious, as we shall see later. In the following account, we explore the power dynamics within the project and some of the tensions that we encountered. We will also reflect on how our approach to collaboration in research affected the goal of inclusion in knowledge production. The account we present is largely based on the experiences of Jonathan and Michaela. It is the account of two members of a larger research team comprised of 17 members who all experienced the project in unique ways.

Power asymmetry and structural inequalities

Jonathan

I was an instructor at the University of Bamenda at the time that I joined this project in 2018. In relation to my colleagues on the project, Michaela Pelican (professor), Alawadi Zelao (associate professor), Deli Terize and Afu Isaiah (lecturers), I was by far the most junior scholar. The positions of lecturer, associate professor and professor in Cameroon are permanent (long term), while that of instructor is short term and very insecure. The University of Bamenda was initially not part of the project, and my role then was limited to mediation between students and the local community in Bamenda, one of the proposed field sites. About a month after my invitation to participate in the project, I suggested the inclusion of my students. I was pleased that no one objected to my suggestion despite the implication that it would require the universities of Yaoundé 1 and Dschang to give up some of the places for their students because the number of students for the project was fixed: six German students and six Cameroonian students.

My late invitation to join the project and the much later decision to allow students from the University of Bamenda to participate in it meant that there was little time for us to prepare: we needed to select students and go through relevant literature. We were not able to properly integrate the project into our study programme as we might have been able to do if we were present at its inception. After a discussion with my head of department and colleagues in Bamenda, we agreed that the participation of our students in the project should count as their internship course – a compulsory requirement for graduation. However, my work with the students faced some obstacles as the University of Bamenda refused to fully support our participation in the project. While my head of department endorsed the project, our application to formally participate in it was rejected by the department responsible for this kind of partnership at the university of Bamenda. It was rejected on the grounds that the selection of the students was not transparent. This was absurd because the applications were reviewed by all five supervisors in the project, and I worked closely with my head of department throughout the process. My head of department refused to give up and decided to let our students participate as interns in the project, a decision that was within his authority. Officially, this meant that I was allowed two visits to Yaoundé to assess the work of the students. In actuality, I needed to be present in Yaoundé throughout the duration of the project and later I was berated for this by an official at the University of Bamenda.

The power asymmetry evident in the above example played out on two levels that affected me and my students in a negative way. As a latecomer to the project, there was very little I could do to alter it, especially in a way that could accommodate some of my needs and those of my students. This is very typical of a collaborative partnership whereby the more powerful actors set the agenda and goals and lead the process. Obviously, this was not the motive of Michaela and her students who throughout the project did their very best to ensure mutual respect and symmetrical collaboration between partners. It was at the institutional level that the power asymmetry played out most noticeably. The University of Cologne, like any other university, was primarily concerned with the interests of its own students and staff, and not necessarily those of the students and staff at the University of Bamenda who participated as partners in the project. Otherwise, a funding requirement for the project would require the collaboration of all partners starting at the level of inception. Internal power asymmetry within the University of Bamenda made my job in the project more difficult by creating additional challenges for me and the students. By rejecting our application to participate in the project, the university failed to acknowledge my supervisory work for the students.

Michaela

Being in charge of managing the project finances put me in a position of power, but this also came with a lot of responsibility and extra work. The project funding was provided by the University of Cologne and required detailed accounting of each and every expense, which posed formidable challenges given the widespread unavailability of receipts in Cameroon. Furthermore, the budget was rather limited, as it usually only covered the expenses of the Cologne students, but in our case was extended to also cover the costs of the Cameroonian collaboration partners. Making ends meet while accommodating unforeseen or changing expenses was a challenging task that absorbed more energy than planned.

Another challenge that affected the power balance involved the regulations and expectations of the partner universities in Cameroon which differed from the status quo in Germany. These included, for example, that in Cameroon academic staff are remunerated separately for all extracurricular activities, including student supervision, extra classes and participation in conferences. Given our limited budget and the different regulations at German universities that count such activities as part of academics' job obligations, I had to navigate a messy middle ground that accommodated both sides' contradictory rules and expectations. In the absence of remuneration for their participation in the project, I had to appeal to my Cameroonian partners to recognise the value of the intrinsic benefit of promoting their students and encourage them to invest in this collaboration with a view to possible future material or intellectual benefits. It worked out in this particular case because we could look back on a long history of collaboration that had produced benefits for all of us in the past. However, there were differences in the commitment of the different partners, which was also owed to the fact that some of us had been collaborating much more closely in the past, while with others, the relationship was rather more instrumental.

On a general note, I acknowledge that all project participants – supervisors and students alike – were deeply committed to the project and to collaborative knowledge production. Also, I wish to acknowledge the significant support provided by the University of Yaoundé 1, which was also the result of the intense lobbying and following-up carried out by my collaboration partners. However, as Jonathan's discussion of his struggle with the university administration in Bamenda indicates, there are often also structural constraints which mean that the success or failure of a collaboration is not necessarily an indicator of the partner's strong or weak commitment to the joint endeavour.

Besides the personal relationship, institutional limitations also weighed heavily on the participants' commitment to the project. As Jonathan explained, the refusal of the University of Bamenda to act as an institutional partner complicated the collaboration. A practical solution was found on the level of the Department for the students to participate in the project by way of internship, which is a study requirement. However, this solution did not cover Jonathan's role as a supervisor in the collaborative research project, resulting in negative consequences, such as reprimands from the university hierarchy. Similarly, due to their different disciplinary background, the students from Dschang were constrained by the requirement to base their MA thesis on their participation in the research project, a limitation which ultimately impacted their motivation and performance. While it was never my intention to disadvantage or exploit some students over others, I realised that student participants from the different universities were unequally positioned in the way they could benefit from their participation in the research project. The majority of the students based in Cologne, Yaoundé and Bamenda were able to write their MA thesis on the basis of the collaborative research, whereas the students from Dschang encountered difficulties in this regard also due to their different disciplinary backgrounds and institutional limitations. However, accounting for these differences was out of my reach, as I had to rely on my Cameroonian partners to mitigate the regulations and risks of their respective institutions.

Facing tension and risk as part of the collaboration

Jonathan

The fieldwork started in Yaoundé with a five-day workshop. We rented accommodation in a gated compound where all the visiting project partners (supervisors and students) stayed for the duration of the programme. The initial arrangement was to provide students with a research allowance to cover fieldwork and living expenses. Under this arrangement, students from Cameroon were responsible for their accommodation. A few days before we convened in Yaoundé, Michaela became convinced that there was enough money to pay for the accommodation of all the visiting students from Bamenda, Cologne, and Dschang, and also that it would facilitate our work if we stayed together. This was good and welcome news for all of us. The students and colleagues from Yaoundé stayed in their own homes but met with us regularly either on campus, at our residence, or in the fieldwork locations.

Four of the supervisors, including Michaela and myself, arrived in Yaoundé about a week before the start of the project and the arrival of the visiting students. We met to make the final arrangements for accommodation, access to the seminar halls, transportation and everything that was needed for a smooth start. It was annoying that our host colleagues did not arrange these things, but I was not surprised because I already knew of the difficulties in Cameroon of getting things done in the absence of key actors and advance payment – therefore completion of the arrangements only became possible with the arrival of Michaela. After the arrival of the students, I found myself in the role of helping German students to navigate their new environment and advising Cameroonian students to be respectful of difference. I slowly stepped into this and other unassigned roles to ensure the best outcome and to avoid conflicts. As we moved forward in the project it became clear that the workload was uneven between the supervisors, not by design but because some were overwhelmed with other obligations outside the project, a factor which meant that they had to delegate their responsibility to others. I ended up supervising more students than the number assigned to me, and so did some other colleagues in the project. We politely discussed this problem a number of times, but very little changed.

My effort to help the students settle in seemed to have had the effect of making them feel comfortable in discussing their concerns with me. They expressed some of their dissatisfaction with the team and suggestions for improvement with me in private. A German student complained about the structure of the workshop, noting that it was more of a lecture, and students were given very little time to contribute to discussions. The student also complained that a Cameroonian supervisor spoke in a sexist and homophobic manner in the workshop. I raised these points with my co-supervisors, and we agreed to caution everyone to be careful with their language and be respectful of difference. Looking at this in hindsight, it seems as if we missed the opportunity to discuss pressing problems in the open and to involve the students in the conversation. But I also realise how difficult it would be to have an open discussion on the issue of sexuality in Cameroon where homosexuality is criminalised.

Students' dissatisfaction in the team was not only directed at the supervisors but extended to fellow students. Two of the German-Cameroonian student tandems did not work well because the students did not have a productive working relation. They resolved the problem by joining different groups. Even among those who got along well, we observed instances when the friendship fractured before eventually getting mended. None of the tensions between students ever got out of control and the supervisors never got directly involved. A few students complained to me about specific issues with their research partner or another student in the group, but they brought the issue to me mainly to seek advice on how to handle the situation. I remember discussing one of those complaints with Michaela after finding out that she heard about the problem from the student who brought it to my attention.

As mentioned earlier, the lack of a forthright and open discussion on discord characterised our approach to conflict within the team. This approach aims at containing and avoiding the escalation of any problem. It is grounded in the idea that conflict is bad and should be avoided at all costs (for a critique of this approach to conflict, see Galtung 2004). This view on conflict is reflected in the dominant approach to intercultural exchanges. Here, respect for cultural sensitivities is paramount. The implication of this approach to our work was that it kept us in our comfort zones, thereby effectively reifying cultural differences within the group and allowing us to take for granted the conventional notions that "otherwise" minorities. In other words, this prevented us from interrogating existing hierarchies that often get in the way of the coproduction of knowledge between actors who occupy different social positions (Aijazi et al. 2021; Zhou and Pilcher 2019).

Michaela

As Homi K. Bhabha, argues, everyone comes to the Third Space with their own background but tries to forget about it in order to create something together (Bhabha 2004). While Jonathan highlights minority-majority dynamics in his conflict assessment, the same situation presented itself quite differently to me due to my role as project coordinator and my positionality as a white female professor.

Given my role as the project coordinator, I felt it was my responsibility to mediate the different expectations and needs of all participants. On the one hand, I noticed the Cologne students' dissatisfaction with the teaching styles of some of the Cameroonian colleagues. On the other hand, I felt it was important to avoid imposing German teaching standards in a project aimed at North-South collaboration; that said, I also took into account the pre-existing power difference which was, in part, effected by the project's funding by the University of Cologne. Another factor that made it difficult to enter into an open discussion was the status and gender hierarchies that play out strongly in the academic context and disadvantage students, female and junior colleagues.[3] Thus, pleading with the students to be accommodating and tolerant of different teaching styles seemed more productive than opening up a critical discussion.

Similarly, dealing with the topic of homophobia was a challenge we had to forgo in this project for the following reasons: Firstly, my colleagues and I did not account for it in the planning and preparation of the project, for example, by including the topic and readings in our preparatory seminar. I believe this would have been an important step toward developing a common understanding, or at least more sensibility for the different perspectives. Secondly, the criminalisation of homosexuality in Cameroon made it a risky subject to discuss in public. However, Jonathan's appeal to supervisors and students to be respectful of difference in our interactions was well received and it facilitated the collaboration.

Project management also entails risk management. Often collaboration with the Global South is more prone to risk than collaboration within established research structures in the Global North. In our project, one of the risks

3 For example, Deli and I (Jonathan) had experienced the effects of these hierarchies in our interaction with members of the university administration when laying the ground for our collaborative project.

was the Anglophone conflict and its effects on the research collaboration. The conflict started in October 2016 as a political crisis, when Anglophones in the North West and South West regions took to the streets to peacefully protest against what has long been perceived as the political oppression of the Anglophone minority by the Francophone majority government. A year later, separatist groups called for the political independence of Anglophone Cameroon (also called Ambazonia) and were met with extreme force by the state military.[4] When planning our academic collaboration, we did not foresee the conflict's violent turn and the resulting insecurity in the Anglophone regions. We responded to the immediate risk by changing the project's regional focus and by integrating colleagues and students from two additional Universities, Bamenda and Dschang. Although the Anglophone conflict was not at the heart of our research project, it affected us in various ways. Several of our project partners, both students and colleagues, were directly confronted with the effects of the conflict: Jonathan and his students came from Bamenda, the capital of the Anglophone North West Region and a stronghold of the conflict. Some project members had family in the conflict region and feared for their safety; others sheltered displaced relatives in their homes. We occasionally talked about the conflict, sometimes in private conversations, sometimes in the group. However, without consciously agreeing to do so, we contained the subject and generally excluded it from our research. In retrospect, this may have been a missed opportunity to learn more about the Anglophone conflict, and how it affected the lives and future making of young people inside and outside the conflict region. At the same time, given different political opinions and the conflict's violent character, it would have been a risky and emotionally challenging subject that demanded more thorough preparation and support than we could provide in this collaboration project.

I agree with Jonathan that by diffusing potential conflict and appealing to mutual tolerance, we avoided descending into the insecurities of collaboration, which according to Zhou and Pilcher (2019) are part and parcel of reaching the Third Space. However, the stated goal of our collaboration was not to reach the Third Space – a concept we did not consciously work with at the time – but to guide students through the process of developing and executing a research project and to promote mutual learning and cultural exchange between project participants. In hindsight, I believe, for project partners to open up to each

4 For a more detailed account of the Anglophone conflict, see Bang and Balgah 2022; Pelican 2022.

other and willingly address negative experiences and divergent opinions, it is necessary that this is jointly agreed on as a common goal and joint way forward. That is, aiming at the Third Space needs to be set in advance as a goal of the project, including the methodological approach it demands. In our collaboration, this was not the case, and thus it is not surprising that we did not achieve or strive for it.

Towards joint production of knowledge in the Third Space: lessons learned

Jonathan and Michaela

This project was grounded in an inclusive and dialogic mode of knowledge production. We took active measures throughout the project to minimise power asymmetry resulting from the different access to resources between German and Cameroonian partners, different positionalities, gender, and age amongst other markers of difference. Although we aimed for a more democratic process of knowledge production and the erosion of established hierarchies in research, our approach did not go far enough in achieving these goals. The main reason for this shortcoming was that our idea of inclusion and collaboration during the project was imprecise and largely informed by general norms of good research practices in anthropology and the social sciences. With the hindsight benefit of reflecting on the project using the conceptual lens of Third Space, we gained a clearer picture of how our approach succeeded, and sometimes failed, in achieving the stated goal of recognition and inclusion of different voices and perspectives in the research process. We discuss some of the successes and failures in the next section.

Starting with access to resources, the major intervention in the project was the decision to distribute the limited financial resources for fieldwork equitably between the German and Cameroonian students. Without this intervention the German students would have received the most substantial share of the research money because it was provided by their home university. This was why the initial distribution of money for fieldwork allocated a smaller amount to Cameroonian students than the sum they received after the intervention. In this case, intervention was spearheaded by Michaela and the German students who were in a position to influence how the funding from the university of Cologne should be used. Given that students worked in pairs comprising of a Cameroonian and German, the lack of the above intervention would have left Cameroonian students with limited financial resources and thus dependent on their German counterparts. Such dependency, as noted earlier, increases the influence of resource rich partners in the research process.

Secondly, allowing students to develop independent studies that were embedded in the main project ensured our goal of symmetrical collaboration. Independent studies sought to answer specific questions central to their aims. They focused on specific groups – students and job seekers, return migrants, cultural or ethnic groups and professionals – and utilised different conceptual and methodological tools that best suited each study. Students working in pairs, which were based on shared research interest, collaborated closely in fieldwork and later in the preliminary analysis of data and presentation of results. They developed question guides and conducted interviews either together, or individually, in a complementary manner. Broader collaboration within the larger group occurred at different stages of the research process. We developed a question guide to help generate data relevant to the general objective of the project. The questions contained in the guide complemented those of participants' individual studies. The results of preliminary findings were first presented and discussed internally, then revised and presented to the general public. However, in this process much of the data analysis was done individually, a point which we shall return to later in the discussion. The project ended with individual reports being produced by students, which they later developed for their bachelor's or master's thesis. The students' reports and the contributions from the supervisors are currently under review for publication in a working paper series at the Department of Social and Cultural Anthropology of the University of Cologne. Our project allowed German and Cameroonian students/colleagues to be fully in charge of their individual research and its outcome and for all the individual researches to contribute to the overall findings of the main project. In this respect, we view the outcome of the project as the result of a joint effort that embraced diverse perspectives and inputs from all partners throughout the research process. This reflects some of the key ideas of knowledge production in the Third Space and our view of research as a collaborative process. It shows how intervention helped to upset the financial advantage of the German partners and the influence that came with it. Intervention also avoided the standard division of labour in North/South that often occurs in research partnerships and limits the role of colleagues in the South to conducting fieldwork while assigning the task of data analysis and writing to colleagues in the North.

Despite our best efforts and the stated achievements, our project did not fully attain the goal of collaboration. The limits of our approach to collaboration as understood through the premise of Third Space were strongest at the level of data analysis and interpretation of results, which constitute a key part of knowledge production. The limitation was because these tasks, as noted above, were largely done individually. Knowledge production in this case was an individual endeavour, which is at odds with the idea of collaboration in knowledge production as the cornerstone of our research partnership. Working individually in this way led to the loss of opportunities for jointly producing knowledge in an inclusive way that embraced diverse perspectives. It spared us the trouble of any disagreement or conflict that could arise from engaging with each other and working across differences. Unfortunately, knowledge production in the Third Space, as explained by Zhou and Pilcher (2019) and Aijazi et al. (2021), is a messy process that is sometimes characterised by conflict. We have already mentioned that our approach to contentious subjects and any form of tension was to de-escalate, often through avoidance. Two students – German and Cameroonian – had very different recollections and interpretations of an interview which they jointly conducted. Jonathan learned about this several months after the students had submitted their reports, and during a discussion of the interview with the German student as a possible case study for her dissertation. He double-checked with the student to make sure that they were talking about the same interview before pointing out the conflicting accounts. The German student was very certain about her account of the narrative and recollection of it. It seemed to Jonathan at the time that one of the students was wrong, but the question was whose account was accurate? Earlier in Yaoundé during our fieldwork, the aforementioned Cameroonian student had shared an observation with Jonathan, saying that during interviews the German student sometimes asked questions that had already been answered, thus implying that she had not understood or paid attention to the interviewee. While this statement suggests that the Cameroonian student might doubt the accuracy of her German partner's account, it also implies that the German student did follow-up questions to get a satisfactory account. Jonathan attributed the different approaches of the students to differences in their backgrounds (disciplinary and cultural among others) and to their stages of academic development. The German student was a master's student in social and cultural anthropology, and the Cameroonian an undergraduate student in development and communication studies. These are only a few possible answers to the question of why these two students gave such different accounts and interpretations of an interview which they had jointly conducted. If the project had established the framework of working in the Third Space, addressing these contradictions would have been a priority, possibly leading to a better understanding of what was said and why it was understood in specific ways and interpreted differently.

Conclusion

We began this article by drawing attention to the structures of exclusion in knowledge production that disadvantage colonised and formerly colonised peoples. Current debates on the decolonisation of higher education address these structural inequalities and call for the recognition of all parties involved and parity in the process of knowledge production. As a way forward towards decolonising the academy and promoting inclusive knowledge production, we have suggested pursuing a collaborative approach that respects a diversity of experiences and perspectives and interrogates existing hierarchies. We have outlined the productive potential of approaching collaboration through Homi Bhaba's concepts of the Third Space, which we understand as a critical moment of intervention that involves descending into the insecurities of collaboration and working through individual, cultural or structural differences. To provide concrete examples of the prospects and challenges of working in the Third Space, we have applied this analytical lens to the collaborative research and teaching project "Urban Youths' perspectives on making a future in Cameroon and beyond" which was realised in 2018 and involved students and scholars from Germany and Cameroon.

Our analysis has shown that achieving the Third Space is a challenging process that requires the commitment and consent of all participants to step out of their comfort zones. Many may not be willing to engage in this process as it can be time-consuming and emotionally taxing. To lay the ground for collaboration in the Third Space, it is thus important to start off with the concept kept clearly in mind and to jointly agree on undertaking the extra effort it will involve.

Furthermore, the analysis of our collaboration project has shown that decolonising higher education is not limited to actively addressing North-South hierarchies, but also acting on power differences (e.g. along the lines of academic status or gender) within university systems. In our project, the focus was on the learning process that students undergo in order to become experienced researchers and work across cultural differences. Even though achieving parity in collaboration was among the project goals, academic hierarchies were maintained throughout the project as a way of efficiently structuring collaboration and sharing responsibilities and tasks. The project coordinators, supervisors and students had different roles and powers, with students supposed to conduct research under the guidance of the supervisors. For those preparing a collaboration project in the future, we recommend considering the following questions as part of the planning phase: Whom should the project benefit and in which ways? Whose collaboration is at the heart of the project (e.g. between students, between supervisors, between students and supervisors)? For which purpose and to what degree can, or should, we break down power hierarchies in research collaboration?

We are confident that collaborating in the Third Space is the way forward to inclusive knowledge production and the decolonising of academia. Descending into the uncertainties of collaboration and dismantling different power hierarchies will help us to recognise and value the experiences and perspectives of those disregarded so far, be they students, minority scholars, or the formerly colonised. More importantly, it will help us to produce inclusive knowledge and good science beyond the narrow confines of Western academia – and this is something which is critically needed to address today's global challenges.

Acknowledgements

We extend our sincere thanks to the students and colleagues who contributed to the collaborative teaching and research project that forms the basis of this article. We are also grateful to the University of Cologne for its generous funding of the project. In developing the analytical framework and writing this article, the authors drew on insights gained from their involvement in a separate collaborative research project entitled "Communication During and After COVID-19", supported by the Volkswagen Foundation (Grant 9699–2).

Bibliography

Aijazi, O./Amburgey, E./Limbu, B./Suji, M./Binks, J./Balaz-Munn, C./Rankin, K./Shneiderman, S. (2021): "The Ethnography of Collaboration: Navigating

Power Relationships in Joint Research." In: Collaborative Anthropologies 13/2, pp. 56–99. https://doi.org/10.1353/cla.2021.0003.

Bang, H. N./Balgah, R. A. (2022): "The Ramification of Cameroon's Anglophone Crisis: Conceptual Analysis of a Looming 'Complex Disaster Emergency'." In: Journal of International Humanitarian Action 7/1, pp. 6. https://doi.org/10.1186/s41018-022-00114-1.

Bhabha, H. K. (1990): "The Third Space: Interview with Homi Bhabha." In: J. Rutherford (ed.), Identity: Community, Culture, Difference, London: Lawrence & Wishart, pp. 207–221.

Bhabha, H. K. (2004): The Location of Culture. (2nd ed.), London: Routledge. https://doi.org/10.4324/9780203820551.

Burke, P./Hsia, R. P. (2007): "Introduction." In: P. Burke/R. P. Hsia (eds.), Cultural Translation in Early Modern Europe, Cambridge University Press, pp. 1–4. https://doi.org/10.1017/CBO9780511497193

Davies, C. A. (1998): Reflexive Ethnography: A Guide to Researching Selves and Others, London: Routledge. https://doi.org/10.4324/9780203069370.

de Sousa Santos, B. (2007): "Beyond Abyssal Thinking: From Global Lines to Ecologies of Knowledges." In: Review (Fernand Braudel Center) 30/1, pp. 45–89.

Fanon, F. (1990): The Wretched of the Earth [French original first pub.1961; first pub. in English in the United States 1965], trans. Constance Farrington, Harmondsworth: Penguin.

Fernandez, C. A. (1992): "La Raza and the Melting Pot: A Comparative Look at Multiethnicity." In: M. P. P. Root (ed.), Racially Mixed People in America, Sage Publ., pp.126–143.

Galtung, J. (2004): Transcend and Transform: An Introduction to Conflict Work, Paradigm Publishers.

Gay Y Blasco, P./De La Cruz Hernández, L. (2012): "Friendship, Anthropology." In: Anthropology and Humanism 37/1, pp. 1–14. https://doi.org/10.1111/j.1548-1409.2012.01104.x.

Ghasemi, P./Sasani, S./Nemati, F. (2018): "Third Space, Hybridity, and Colonial Mimicry in Fugard's Blood Knot." In: Khazar Journal of Humanities and Social Sciences 21/1, pp. 34–52. https://doi.org/10.5782/2223-2621.2018.21.1.34.

Hall, S. (1992): "The West and the Rest: Discourse and Power." In: Stuart Hall/Bram Gieben, Formations of Modernity, Cambridge: Polity Press, pp. 275–332.

Holliday, A./Amadasi, S. (2019): Making Sense of the Intercultural: Finding De-Centred Threads, London: Routledge. https://doi.org/10.4324/9781351059190.

Hountondji, P. J. (2004): "Knowledge as a Development Issue." In: K. Wiredu (ed.), A Companion to African Philosophy (1st ed.), Blackwell Publishing Ltd, pp. 529–537. https://doi.org/10.1002/9780470997154.ch45.

Howell, P. (1996): "Book Reviews: The Location of Culture. By K. Homi Bhabha. London: Routledge. 1994. xiii + 285 pp. £11.99 paper. ISBN 0 415 05406 0." In: Ecumene, 3/1, pp. 113–117. https://doi.org/10.1177/147447409600300117.

Khosravi, S. (2007): "The 'Illegal' Traveller: An Auto-ethnography of Borders*: THE 'ILLEGAL' TRAVELLER." In: Social Anthropology/Anthropologie Sociale 15/3, pp. 321–334. https://doi.org/10.1111/j.0964-0282.2007.00019.x.

Martin, S. B./Dandekar, D. (eds.). (2021): Global South Scholars in the Western Academy: Harnessing Unique Experiences, Knowledges, and Positionality in the Third Space, London: Routledge. https://doi.org/10.4324/9781003109808.

Mbembe, J.-A. (2016): "Decolonizing the University: New directions." In: Arts and Humanities in Higher Education, 15/1, pp. 29–45. https://doi.org/10.1177/1474022215618513.

Mitlin, D./Thompson, J. (1995): "Participatory Approaches in Urban Areas: Strengthening Civil Society or Reinforcing the Status Quo?" In: Environment and Urbanization 7/1, pp. 231–250. https://doi.org/10.1177/095624789500700113.

Ndlovu-Gatsheni, S. J. (2018): Epistemic Freedom in Africa: Deprovincialization and Decolonization. (1st ed.) London: Routledge. https://doi.org/10.4324/9780429492204.

Ngeh, J. (2021): "Inclusion and Exclusion of Postcolonial Subjects in Knowledge Production. Academic Experience in Sweden, Cameroon, and Germany." In: S. B. Martin/D. Dandekar (eds.), Global South Scholars in the Western Academy (1st ed.), London: Routledge, pp. 65–76 https://doi.org/10.4324/9781003109808-8.

Ngũgĩ wa Thiong'o (1986). Decolonizing the Mind: The Politics of Language in African Literature. Pourtsmuth N. H. Heinemann.

Nyamnjoh, F. B. (2012): 'Potted Plants in Greenhouses': A Critical Reflection on the Resilience of Colonial Education in Africa." In: Journal of Asian and African Studies 47/2, pp. 129–154. https://doi.org/10.1177/0021909611417240.

Pelican, M. (2022): "The Anglophone Conflict in Cameroon – Historical and Political Background. Arnold-Bergstraesser-Institut." In: ABI Working Paper No. 20. https://www.arnold-bergstraesser.de/leistungskatalog/the-anglophone-conflict-in-cameroon-historical-and-political-background (Accessed 03.10.2023).

Said, E. W. (1979): Orientalism (1st Vintage Books ed.), Vintage Books.

Scholte, B. (1974): "Toward a Reflexive and Critical Anthropology." In: D. H. Hymes (ed.), Reinventing Anthropology (2nd ed.), New York: Vintage Books. pp. 430–458.

Smith, L. T. (1999): Decolonizing Methodologies: Research and Indigenous Peoples. (1st ed.) London: Zed Books Ltd.

UNESCO Institute for Statistics. (2015): "How Much Does Your Country Invest in R&D? UNESCO Institute for Statistics [dataset]." Do you want to cite this Blogpost from 2016: https://uis.unesco.org/en/blog/how-much-does-your-country-invest-research-and-development-rd (accessed 03.10.2023)

Wade, P. (2004): "Images of Latin American Mestizaje and the Politics of Comparison." Bulletin of Latin American Research 23/3, pp. 355–366. https://doi.org/10.1111/j.0261-3050.2004.00113.x.

Walter, E. A./Jerome, A./Marceline, D. N./Yakum, M. N./Pierre, W. (2017): "Map of Biomedical Research in Cameroon; a Documentary Review of Approved Protocols from 1997 to 2012." In: Globalization and Health 13/85, pp. 1–8. https://doi.org/10.1186/s12992-017-0312-y.

Wolf, M. (2000): "The Third Space in Postcolonial Representation." In: S. Simon/P. St-Pierre (eds.), Changing the Terms. Translating in the Postcolonial Era, University of Ottawa Press, pp. 127–146. https://doi.org/10.2307/j.ctt1ckpcz7.10.

Zhou, V. X./Pilcher, N. (2019): „Revisiting the 'Third Space' in Language and Intercultural Studies." Language and Intercultural Communication 19/1, pp. 1–8. https://doi.org/10.1080/14708477.2018.1553363.

Doing Anthropology at Home, in Chad

Remadji Hoinathy, Djimet Seli and Andrea Behrends

Introduction

The three authors of this paper are situated in academic and institutional contexts in Chad and Germany, and each of us is involved in international academic and research contexts in Europe and Africa and beyond. Our collaborations reach back more than fifteen years to the time when we inaugurated the *Centre de Recherches en Anthropologie et Sciences Humaines* (CRASH) together with researchers from Chad, Europe and the US. In this paper we connect the decolonisation paradigm to practices of collaboration in research projects based on our personal experiences together and with other research partners. Our focus lies therefore on collaborations between so-called Global South and so-called Global North[1] researchers, but it also goes beyond such collaborations to look at the situation of anthropology in Chad, specifically. We do so because we think that the ways CRASH deals with the multiple contextual constraints it is facing can illuminate how and why we need to develop a decolonising perspective on methods, outcomes and avenues of anthropological research, with a particular focus on the possibilities for supporting young researchers.

1 We are aware that the use of the terms "Global South" and "Global North" is problematic in attributing vague spatialities to globally unequal access and regulation of funding and research (as in the case of this article). Following Ndlovu-Gatsheni (personal conversation), we assume "Global South-related thinking" to also happen in what is perceived as the "Global North", and vice versa. The connection of geographic spheres through a coloniality of suppression and extraction cannot be put into simple dichotomies but needs to be analysed in the complex entanglements that colonial regimes and their aftermath produced. In referring to these notions in this paper, we want to allude to the continuing existence of power differences and postcolonial dependencies, but to avoid drawing clear boundaries about where such differences and dependencies are currently produced.

Constraints range from lack of funding for local researchers, to the need for producing applied research in consultancies, and to the lack of appreciation of such work in global academic circles. The Chadian researchers have found solutions to navigate these constraints – but they lack full appreciation of their work by academic institutions in the Global North. We maintain that this lack of appreciation is a result of colonially based forms of knowledge production and dissemination that needs to be revisited in order to establish new forms of valuing academic work.

To decolonise involves looking at all aspects of academic knowledge production and use, such as research funding, data collection, analysis, writing and publication of findings. It also requires actively engaging with methods in ways that do *not* reproduce existing power differences that become visible not only through the direction of money flows within project work, but also through forms of project management. Our contribution takes a closer look at the uses of project results, and the way they come to be valued by all project partners, but particularly by the funding agencies. We maintain that collaboration between scholars from African and European countries is of great value when proposing projects: this applies to the extent where a project that is devoid of such established or aimed-for partnerships and collaborative methodologies lacks credibility and risks failure to secure funding. This focus does not, however, give any information about the value and later use of the collected data. We maintain that while European, and particularly German, funding agencies rate such collaboration very highly, the usual expectation is that contributions from collaborative projects will later enter academic debates that are specific to European and Northern American interests. Similarly, ranking practices according to the citation indexes that are applied to publications arising from collaborative project work, result in a preference for dissemination in prestigious journals mainly produced in the Global North. Non-academic outcomes such as applied anthropology that aims at producing justice in relation to inequalities of access and contribution – often within the country where research has been undertaken – do not attract much appreciation in European or Global North academic institutions and in project evaluations conducted by funding agencies. We argue for the establishment of an appreciation of various, and independent, forms of usage of research data – or, in other words, to decolonise anthropology through acknowledging that the value of data is not exclusively produced through highly rated (Global North) conference presentations, journal or book publications. In this paper we promote the idea that considering and appreciating different forms and means of using research data are equally important. Such independence is of considerable relevance in any effort at decolonisation. After all, decolonisation is an abstract term and only the practice of producing and using data can invest the project of decolonisation with practical and meaningful life.

This connects to the realisation, that even if individual collaboration practices are as symmetrical as possible under the given structural constraints, the actual necessities and configurations around social science research are not equal. Research funding is usually low in most African contexts, and the need to produce data "on demand" for applied purposes is not always a researcher's primary goal, but more often a means to survive professionally (Behrends 2015). In our example, we look at the workings of an anthropological research centre in Chad, which has to navigate the demands of both academic and consultancy work – with the declared aim of rendering academic careers possible for anthropology students who have attained their MA degree at the University of N'Djamena, in Chad's capital city. We maintain that the recent and much wider acceptance of "applied" or "activist" research in European contexts, as well as the ever-growing incentive of forging symmetrical collaboration between researchers from the Global South and the Global North, has changed the Western, formerly negative, perspective concerning the applicability of academic research results. We begin by giving an example from our own experience to demonstrate the different perceptions of project work. Following this, we consider the prospects for the establishment of a specific Chadian anthropology by discussing the particular constraints and challenges affecting the discipline in the Chadian context. We outline a preliminary solution for overcoming these challenges. In our conclusion we revisit the idea of independent and equally valued research contributions and assess their prospective uses.

To illustrate different understandings of the outcomes a project might produce, we start with an example from our own experience. From 2006 to 2011 Remadji Hoinathy (as PhD candidate) and Andrea Behrends (as project coordinator) collaborated in a project with the title "Travelling Models in Conflict Management", funded by the Volkswagen Foundation. The project focused on how ideas concerning the "good" management of conflicts that were developed in one place become translated through human and non-human mediators and are subsequently received, and often changed, in another. In this project six German researchers and six academics from various African countries participated, supervising together a group of seven PhD students from six African countries. Why do we begin with this memory of a project? The point we want to make is about varying expectations directed at the outcomes of research.

Looking at models to manage conflict was, from one point of view, an exercise in understanding translation processes (Behrends et al. 2014; Hoinathy and Behrends 2014; Rottenburg 2009). How do ideas develop? How are they transferred and put to use in another context? Whose interests, whose knowledges are relevant – and how do contingency, politics, and the situatedness of actors interfere in the process of transmission? In most of the individual research projects (about, for instance, "power sharing", "mediation through elders", "freedom of speech", "neighbourhood watch" or "fair revenue distribution") the result was disillusioning: most of the models did not reap the anticipated results. Most did "not work". However, according to the project's setup, the models' usefulness was secondary. But we did have discussions about exactly this point. Why do we study conflict intervention if we do not want to make models better? If we find out how they fail to achieve what they were originally designed for, why not suggest ways of making them more understandable? Why waste the energy and the financial means to develop conflict intervention models, study them, and then come out with the same result? Namely, that they often do not work. Full stop. We found that the answer to the question of *why* a model of conflict intervention does not work was considered secondary or even unimportant to the question of *how* it does not work. The second question was considered academically relevant, while the first one was considered as "applied research", with a tint of development aid or policy orientation. Our contribution to this book not only revisits such assumptions of relevance and hierarchy but seeks to actively argue for a multiplicity of outcomes and forms for academic work that would make it possible to decolonise by giving room to creativity and context-relevant adaptations of our work so that it can receive the appreciation it deserves.

This paper is an adapted version of a presentation by Remadji Hoinathy on the situation of anthropological teaching and research in Chad, a country where the subject has only very recently been introduced, in 2014. Djimet Seli and Andrea Behrends have, in different ways, contributed to the paper – Seli providing background information on the development of the university institute of anthropology at the University of N'Djamena, and Behrends on the collaboration between the Chadian anthropological research institution, CRASH,[2] and international partners. In general, the chapter aims at contribut-

2 The *Centre de Recherches en Anthropologie et Sciences Humaines* (CRASH) has been established as an internationally accessible research centre. It is led exclusively by Chadian anthropologists but has an international board of directors.

ing to the extension of Law's and Lin's (2017) suggestion that it is time for anthropologists to "provincialise anthropology" in the way the Dipesh Chakrabaty (2000) proposed to "provincialise Europe" – a source to whom the two authors explicitly refer. While Law and Lin write about the different intentions and uses of anthropological research in a European (or more explicitly, British) and in a Chinese context – so quite close to our intentions in this chapter – Chakrabaty addresses the particular difficulties of Western trained Southern anthropologists whose awareness of actual asymmetries of knowledge makes them deeply uneasy when considering the different forms of knowledge they should prioritise (Law and Lin 2017: 3). Anthropology (and Science and Technology Studies), so Law and Lin maintain, would need a "postcolonial version of the principal of symmetry", one in which "the traffic" between different locations "would be lively, two-way, and contested" (Law and Lin 2017: 4). Their approach speaks to the story we started out with, as it looks at whose priorities count, and how to deal with differential requirements geared towards academic research – without valuing one higher than the other. To guarantee symmetrical collaboration, for instance, Law and Lin underline that besides empirical research and theorising, "some pretty matter-of-fact institutional practicalities" (Law and Lin 2017: 7) should be well thought about. This coincides with our objective of seeing what chances a Chadian anthropology has of being heard and introducing changes worldwide. Ideally, this venture would avoid the strengthening of national boundaries precisely because it is directed towards overcoming barriers. It is, however, necessary to understand historical, current and future oriented perspectives on anthropology in the history of Chad and to realise how these factors influence not only the possible impact of studies, but also the demand for anthropologically derived knowledge. Importantly, particular attention needs to be paid to the practicalities of conducting research in a place where anthropology is (to put it mildly) not a discipline held in the highest national regard. The question we ask in this paper is about the possibilities of defining an anthropology that covers matters of concern within the Chadian context and, at the same time, aligns with international academic fora. We argue for the need for a model forged via practice and the contingencies associated with that practice in the Chadian context. Prospectively this model would be capable of "de-stigmatising" the combination of consultancy (applied research) and academia (considered to be "fundamental" research). Rather than turning their backs on each other, these practices can mutually enrich each other, as the turn to "public anthropology" has also shown in recent years.[3] We will first set the scene of anthropology's history in the Chadian context and then move on to describe its present situation and its possible (that is, desired) future.

Anthropology has been one of the latest subjects to be introduced into the Chadian higher education curriculum and was achieved with support from the University of Aix en Provence (France). This late entry, in 2014, reflects the perception of the discipline and its importance in Chad. Indeed, compared to other countries in the Central African sub-region, Chad has been the last one to introduce anthropology into the academic curriculum, and exclusively at the University of N'Djamena, in Chad's capital. Creating public interest for this discipline is therefore far from being achieved in a country where, in general, academic research, especially in the humanities, is very undervalued. In addition, Chad is among the countries least visited by foreign researchers since successive periods of dictatorship and deprivation of liberty have not encouraged the emergence of neutral and free scientific activity. In such a context, anthropology in general – but especially for Chadians (i.e. "at home") – is indeed a real challenge. The process of change is a winding course dotted with many obstacles. However, it is also a journey full of opportunities to be seized. How to stay on course despite a great number of difficulties and how to seize opportunities without going astray is, thus, the question.

This paper is based on the authors' experiences and professional trajectories, but it also describes the experiences of other anthropologists who have worked in Chad or elsewhere. This experience has been developed over more than a decade of what might justifiably be called an "academic adventure" in the specific context of Chad. That said, the "adventure" has always been undertaken within a context of close international academic collaboration – a feature which played a key role in this particular experience. The three authors have followed different pathways in their anthropological careers. Remadji Hoinathy pursued an MA in social sciences at the Catholic University of Central Africa in Yaoundé, Cameroon, before embarking on a PhD at Martin Luther University of Halle-Wittenberg, funded by the Max Planck Institute of Social Anthropology. Djimet Seli received a scholarship from the University of Leiden for his PhD, after having studied history and communication at

3 See, for instance, the Public Anthropology Working Group of the German Anthropological Association, which was founded in 2021 and has as its mission to "practice publicly accessible, intersectional and decolonized anthropology" (http://publicanthropology.de/).

the University of N'Djamena and anthropology at the Yaoundé I University. Andrea Behrends studied anthropology at the Free University in Berlin, where she also obtained her doctorate. She is now professor for social and cultural anthropology at Leipzig University.

Setting the scene

Chad, an unstable African hinterland

Located in Central Africa, Chad is a vast area of 1,284,000 km2, covered in its northern part by the Sahara and the Sahel regions. It is inhabited by about 16 million people, of which 52% are women, and 50.6% are young people. Most of the population live in rural areas and depend on the agro-pastoral sector. The area-to-population ratio gives a density of 13 inhabitants per square kilometer. It is therefore a sparsely populated country, where most people live in the central and southern parts. Politically, Chad is known for instability, particularly after the country's independence from French colonial rule in the 1960s, with long periods of civil war and other crises. The country has experienced successive authoritarian political regimes. Despite the official start of a democratic era in 1990, freedom and democracy are still far from being guaranteed. Economically, the country, for a long time, depended mainly on agriculture (especially cotton) and livestock (with a livestock population of at least 100 million head), before it embarked on producing crude oil in 2003. From then on, the country's economy has almost exclusively depended on this particular resource (Hoinathy 2012; Hoinathy and Behrends 2014; Behrends and Hoinathy 2017). The chaotic governance of oil resources (Hoinathy and Janszky 2017), combined with the rapid decline of oil prices from 2014, has led the country into its current and, in this form, unprecedented economic and social crisis, drastically reducing the state's ability to meet social demands and household purchasing power.

Chad is also undergoing a difficult socio-political situation: the population suffers from the plights of Islamist extremism, the presence of armed gangs and the recurrence of rebellions in the north of the country. In April 2022, President Deby (who had been in power for 31 years) was killed in fighting between the national army and the rebel group Front pour l'Alternance et la Concorde au Tchad (FACT). His son, Mahamat Deby, has taken over the leadership of the country and of a transitional military council, a body which is hardly in a position to bring the country back to constitutional order (Hoinathy 2022). Currently, inter-community conflicts are also rising in a large part of the country's territory. Chad's central location puts it at the crossroads of the political and security dynamics at work in neighbouring countries. Like other Sahel countries, Chad has, for more than a decade suffered the consequences of the highly volatile situation in Libya, resulting in chronic instability in its northern part. On the western front, Chad, along with the other countries bordering Lake Chad, is suffering the wrath of factions of the violent extremist group Boko Haram. In the south, the Central African Republic's lack of control over its northern part is also fuelling dangers on Chadian soil. Finally, the situation in Sudan poses huge humanitarian and security risks (Hoinathy and Yamingué 2023).

One single university, then several, but without clear funding

In Central Africa, excepting Cameroon, which created the University of Yaoundé in 1963 (three years after independence), the other former French colonies did not set up universities until much later: the Central African Republic, with the University of Bangui in 1969, Gabon, with the University of Libreville in 1970, and then the Congo and Chad, with the Universities of Brazzaville and N'Djamena respectively in 1971. As with these other countries, Chadian officials had to train mainly in France, but also in Soviet universities and later in the Middle East (Ramoupi/Ntongwe 2016).

For several decades, higher education exclusively meant the University of Chad,[4] located in the country's capital, N'Djamena. Inaugurated in 1971, this university aimed to educate young Chadians who aspired to be leaders in the development and governance of the state. It was at this time that Chadian university teaching was first implemented. During the colonial era, Chadians went to study at the Ecole Normale William Ponty in Dakar, Senegal, or in Brazzaville, the capital of French Equatorial Africa. After independence, Chadians also went to France as a preferred destination for higher education. Initially, the University of N'Djamena consisted of only two campuses: 1) Ardep Djoumal, dedicated to humanities, management, and law; and 2) Farcha, responsible for instruction in the "hard" sciences. From the 2000s onwards, other universities and institutes opened, bringing the number to eight different universities located in the country's seven major cities. These

4 Turned into University of N'Djamena at the beginning of the 2000s.

are N'Djaména, Abéché, Moundou, Sarh, Doba, Mongo, Ati and Pala. Apart from N'Djaména, Abéché, Moundou and Sarh, the other towns have relatively little basic infrastructure and remain relatively rural. The oil boom of the 2000s made it possible to start investments, but unfortunately these proved to be inadequate for raising the level of development of these cities.

In 2012, the government also opened a third campus in Toukra, on the outskirts of N'Djamena. In addition to these universities and campuses, seven institutes of higher learning and four teacher-training colleges are under the administration of the Ministry of Higher Education, Academic Research and Innovation in Chad.

The King Faisal University is one of the first private universities in Chad to promote scientific research and innovation. Created in 1992, it teaches in Arabic only. Originally housed in the central mosque of N'Djaména, it later had its own campus and obtained the status of a public university under governmental supervision. The other private universities are notably the Emi Koussi University, the Université la Francophonie, the Université Hautes Etudes Commerciales-Tchad and the Université Africaine de Management et de L'Innovation, all based in N'Djaména.

The quantitative, but above all, the qualitative challenges of higher education in Chad

Collectively Chad's universities teach about 57,443 students (Alwhilda Info 2023). Each year, only 25,000 students are accepted at universities to receive higher education – a number that is far below the actual demand, and largely composed of young men. Indeed, women's enrolment remains lower than that of men and, in addition, the dropout rate of women is higher, reducing their numbers in higher education even more in comparison to men.

In addition to the challenges related to the inability of the Chadian university system to meet the demands of young Chadians, the quality of teaching and research remains equally deplorable for the following reasons.

- The quantitative and qualitative insufficiency of human resources: all eight universities share about 1,346 staff, the majority of whom (68%) hold a master's degree or other equivalent degrees. In recent years, very significant efforts have been made not only in terms of increasing the number of new doctorates, but also in the advancement of the capacities of the African and Malagasy Council for Higher Education (CAMES). This implies a greater number of teaching staff authorised to supervise students enrolled in the master's and doctoral cycles.

- Insufficient financial and material resources (infrastructure, equipment, research laboratories, etc.): by 2022, the budget of the whole Ministry of Higher Education was approximately 2.36% of the national budget.
- Teaching conditions: it is generally observed that the indicators relating to teaching conditions are not favourable. Some of the institutions, especially in the provinces, do not have their own rooms but work in rented premises.
- Low use of ICTs in teaching and research: Internet access remains expensive despite the latest price cuts. At the same time, none of the campuses have free access for students and teachers.
- Poorly supported research characterised by insufficient infrastructure and a lack of a relevant programme oriented towards sustainable development. The government is developing relatively little funding to enable sustained research. In fact, there is no consistent and permanent mechanism for funding research projects carried out by teacher-researchers at the various universities, which necessitates searching for research funding outside the national context. Consultancy work is one avenue that allows Chadian researchers to pursue research work, however, not always in the direction of their own interests. Based on this necessity of outside dependence, Hoinathy has developed a plan to give young researchers a chance to qualify not only in good consultancy, but also academically. We develop this idea below.

Anthropology in Chad, like sowing seed in a dry land

The latest subject taught at the University of N'Djaména

For a very long time, the social sciences were ignored in higher education in Chad. Sociology made its entry in 2003, but it was more than 10 years later that anthropology was introduced, following an initiative of the late French professor Bruno Martinelli. His proposal to open a new and independent anthropology department that would rely on French financing and technical support from the University of Aix-en-Provence was at first heavily disputed and refused by the sociologists at the University of N'Djamena who wanted to graft anthropology onto their own department. Although the department has worked quite well, with the support of professors from the Universities of Aix-en-Provence, Leiden, Halle and elsewhere, the university's administration has continued to keep a suspicious eye on the discipline.

When it first opened, the Department of Anthropology only taught up to the levels of Master 1 and 2. On the basis of an application and a competitive examination, this master's degree programme accepted students with a bachelor's degree in the social sciences and humanities in general (sociology, geography, history, linguistics, philosophy, law, literature etc.). For the last three years up to 2022, the bachelor's degree cycle has been introduced in addition to the master's course. Master's students come from all disciplines of the humanities and social sciences, but mainly from sociology.

The main subjects taught at the department of anthropology are:

- Social Anthropology of Development
- Anthropology of Communication
- Anthropology of Kinship
- Social and Cultural Anthropology
- Legal Anthropology
- Natural Resources and Governance
- General Ethics
- Research Methodology
- Project Evaluation
- Epistemology of Anthropology
- Anthropology of Religion
- Anthropology of Health
- Anthropology of Organisations
- Anthropology of Gender
- Economic Anthropology
- History and Purpose of Anthropology
- General Sociology
- Academic Work, Pastoralism
- Major Anthropological Theories
- Anthropology of Chad
- Anthropology of Central Africa etc.

The curriculum is inspired by the Cameroonian and French models. The staff consists of a total of 12 members: ten nationals and two foreigners. Among them, there is only one woman. There are eight PhD holders and four master's degree graduates.

Anthropology is a subject that young Chadians have embraced as an opportunity to not only study their own country and see it from new perspectives, but also to learn about methods of engaging with pressing issues. Having started with 17 master's students, the department has expanded to include about 450 students working at bachelor and master levels. To date, over thirty anthropologists at master's level, (four women and 29 men), have already defended their thesis on topics as varied as:

- The Chadian prison system (anthropology of bureaucracy)
- The governance of petroleum resources through global transparency initiatives
- The challenges for health care provision in rural areas
- The question of food and nutrition
- The management of natural resources around Lake Chad
- The dynamics of livestock farming and pastoralism

Some of these students have entered foreign universities to pursue a doctorate. Other former students have found work in the development and humanitarian sectors, and some have joined CRASH. Among those who joined CRASH, one defended his PhD thesis in France in 2022, while three more are still in the process of completing their research degree. Other graduates are engaged in making applications for research projects and / or in search of funding. In general, a demand for anthropological expertise and research obviously exists in Chad and the usefulness of well-trained researchers is well recognised, especially by the development community and humanitarian organisations.

Although the initiative for the creation of the department came from Professor Martinelli with the support of French cooperation in Chad, Chadian teachers were, from the outset, deeply involved in curriculum development, thus ensuring that realities and needs at the Chadian level were considered in terms of both teaching and research content. Importantly, the intention of creating the department was to provide an opportunity to train a new generation of researchers able to investigate, report on and understand the complexity of the local region in relation to the global situation, thereby strengthening the capacities of national research and development organisations. As the department's staff also have the responsibility of supervising students for their research, they contribute to the training of autonomous and emancipated Chadian anthropologists. Despite these features a persistent problem remains, namely the lack of means for enabling these anthropologists to have an autonomous research agenda.

With the opening of a bachelor's degree in 2019, the department ventured to hire more students and, above all, to better supervise them from the first year of their university study. This change offers the future prospect of recruiting young graduates who have already defended their master's degree and embarked on careers as lecturers.

Little known and not always understood

"What do you do for a living? Anthropologist? What is that now?"
"Anthropologist? Okay, you're the one who discovered Toumai?" (a spectacular archaeological discovery).
"Anthropologist? Okay, you're the ones looking for the skulls here?"

This is a non-exhaustive list of questions that all colleagues at the department have had to answer each time we introduce ourselves as anthropologists. Even when people seem to have heard of anthropology, they understand it as a backward science, locked in the past and focused on the study of the cultures and customs of so-called traditional societies, as was the case under colonisation. The fact that anthropology is the academic discipline that captures the dynamics of our current societies and therefore advances with them is poorly recognised. Hence questions such as, "But what does an anthropologist do in researching about oil?" – imply that the researcher should instead be in a remote village studying customs. Anthropologists face this common misunderstanding not only in Chad, but everywhere. We argue that the public notion of anthropology being the discipline that preserves supposedly "traditional" cultural values originates in the discipline's origins as a colonial endeavour with a canon built on the experiences of (mainly) men who travelled around the world to "discover" cultures. Regarding anthropology as a discipline that actively engages society and its various "matters of concern" (Latour 2004) is only slowly entering public understandings, if at all.

These misperceptions mean that, in general, anthropology remains an academic discipline that is not very popular in Chad since the subject's usefulness and relevance are not obvious to decision-makers or the general public. As anthropology's use as a fully-fledged scientific discipline capable of contributing to the country's development is not clear to the authorities, they continue to be reluctant to devote a place to it in the study programme. Compared to sociology, it is true that anthropology has struggled to establish itself in many other African countries. Onyango-Ouma (2006) clearly documents how in Kenya, a president who is himself an anthropologist, has been reluctant to support the opening of a department or faculty. Chad is the last country in the sub-region to open its doors to the discipline. When we see the small amount of interest that the international academic community has so far taken in anthropological studies on Chad, we also realise that the country has not been the most visited one in the sub-region.

The problem of limited international interest

As in most other landlocked former French colonies, anthropological work in Chad was first carried out by colonial missionaries and administrators and researchers from institutions dedicated to the study of overseas societies and later by the French *Centre National de Recherche Scientifique* (CNRS). Currently the number of international researchers working on Chad is still very limited: consequently publications about Chad are sparse. Factors including long periods of civil war, successive dictatorships and the control they exercised over the knowledge community have obstructed any significant revival of the interest of foreign anthropologists in the country. In the mid-1990s, some international anthropologists cautiously returned to Chad, but in far fewer numbers compared with other countries of the larger region.

Regarding Chadian anthropologists, some, like Remadji Hoinathy and Djimet Seli who are two of this article's authors, have been trained in neighbouring countries, in West Africa and also in Europe. To date, there are still only three Chadian anthropologists who hold a PhD, a figure that clearly indicates the extent of the task to be covered not only in terms of teaching, but also in the production of independent and critical scientific knowledge, and in action research for larger scale political decision-making.

The adventure of CRASH

When the French anthropologist Bruno Martinelli helped to open up the anthropology department in 2014, a professor at the University of N'Djamena advised him to turn to the only anthropologists of the country, that is, to the researchers at CRASH who then became the first to teach in the new Department of Anthropology. Following this, CRASH put its full energy into building up and running the university institute. But before we speak of the university institute, we want to outline the history of the centre. CRASH is a multidisciplinary research centre based on anthropological methodologies that aims to promote research in social sciences and humanities for development purposes. It is a partnership between Chadian and foreign researchers who all work in and on Chad, or the Lake Chad Basin, and who are committed to supporting and advancing academic research in the country. Thanks to this partnership the centre came into being in 2005 and was officially recognised by the Chadian Ministry of Interior and Public Security on 11th September, 2007 under folio No.2696 as a private institution outside the university.

Thanks to the support of its international partners such as the Max Planck Institute of Social Anthropology (Halle, Germany), the African Studies Centre (Leiden, Netherlands) and the University of Bayreuth (Germany), the centre has gained increasing visibility and established a solid reputation in fundamental and action research both locally and internationally. This opened space for new partnerships and scientific collaborations. Between 2005 and 2023 CRASH has achieved the following:

- CRASH researchers carried out a large number of international collaborative research projects;
- They gained a number of international excellence scholarships and fellowships;
- They conducted and successfully finalised several local and regional action research projects for local and international development agencies;
- CRASH trained a group of Chadian researchers: three PhDs and, three PhD candidates and eight master's degree students;
- CRASH integrated into international research networks;
- CRASH researchers have published in numerous peer-reviewed books and journals;
- They set up a multidisciplinary documentary database on the various issues covered by the centre and its experts and a website (http://www.crash-td.net);
- CRASH researchers contribute essentially to the Department of Anthropology at N'Djamena University through lecturing, supervision and offering internships to students.

The centre thus established anthropology and has been practicing it since 2008, when Hoinathy and Seli joined it with the dual agenda of doing a doctoral thesis on one side and developing the centre on the other. Initially Remadji Hoinathy worked on natural resources, especially oil, and then moved on to cover a variety of topics such as security, migration, and civil society. Djimet Seli's research first focused on the impact of mobile telephony in relation to the connections between diasporas and their locality of origin (Seli 2014). He then focused on the dynamics of pastoralism, migration, gender-based violence and conflict. Basic research has been made possible through collaboration with international academics, including collaborative projects with the Universities of Halle and Leiden. The Volkswagen Foundation's Excellence Scholarship has been a major contribution to the development of Remadji Hoinathy's work. Andrea Behrends has been part of the centre's advisory board from the beginning. Her research on oil and social change in the south of the country as well as on displacement, emplacement and aid, particularly in eastern Chad, was supported by the Max Planck Institute for Social Anthropology, the Volkswagen Foundation and the German Research Foundation. Importantly, CRASH's staff and infrastructures provided her with documents to be able to travel and do research in the country, and she continues to engage with the centre and its researchers.

While the university and public policy makers in Chad were slow to realise the importance of the work conducted within CRASH, international collaboration has made it possible to establish a solid reputation for CRASH and its researchers. This has resulted in progressive and regular solicitations from development and humanitarian action communities for action research. Thus, researchers working both within CRASH and individually, have conducted more than twenty studies on topics as varied as:

- Migration
- Violent extremism
- Public Health
- Intercommunity conflicts and cohabitation
- Security governance
- Climate Change
- Evaluations of various development projects

These studies were commissioned by various United Nations bodies including the United Nations Population Fund (UNFPA), the United Nations Development Programme (UNDP), and the International Organization for Migration (IOM). Further commissions came from the European Union or other financial partners such as SWISSAID, the *Deutsche Gesellschaft für Internationale Zusammenarbeit* (GIZ), and national NGOs.

Implications and challenges

We now would like to address lessons learned from such development, as despite the very successful trajectory we have described, anthropologists in Chad continue to face a number of challenges.

Reflexivity and the challenge to remain neutral

At the epistemological and ethical level, the question arises of the ability to work on questions and problems that are directly related to the researcher's own history and daily life. Because, most evidently, whatever competences are possessed by the individual researcher, they are always speaking from a specific position. The very nature of the subjects and issues studied can add to the precariousness of the anthropologist's work. Let us contextualise this challenge with an example: a study about social changes brought about by the oil industry in rural areas of southern Chad. Knowing that oil has been the essential resource of the Chadian State, authorities, communities and citizens (including the researcher) expected oil to be *the* factor for development, modernisation, and general wellbeing. Social science research on oil or mining has documented the great aspirations that these strategic resources created in the contexts where they have either been discovered and/or prospects for exploitation were announced (e.g., Coronil 1997; Apter 2005; Fergusson 2005; Reyna and Behrends 2011).

For Chadian researcher[5], studying this question is both exciting and risky. Indeed, at the same time as the researcher is making observations on the ground, he has also been observed and his movements more or less controlled through the obligation to pass through the office of the regional governor, the prefect, the sub-prefect, the cantonal chiefs and their subordinates, every time he ventured out to a new locality. He had to deal with this reality until the end of the field research, taking precautions such as always traveling with official

5 As the next paragraphs on researching oil in southern Chad has been written (and the research has been carried out) by Remadji Hoinathy, the masculine pronoun is used because the description gives a first-hand account of his experience.

authorisation, discreetly using his field equipment (the camera in particular), having a representative of the canton chief accompany him on long journeys in case he would meet an oil security patrol. In such a context, making the anthropology of oil "at home" makes contact with the populations in the field easier, but it also exposes the researcher to sometimes very restrictive control in the field, thanks to a better understanding of the context and less of a linguistic barrier.

Controversial debates that began at the beginning of the negotiations on the Chad-Cameroon oil project have made the relationship between the actors involved in the issue quite complex. In Chadian public opinion, the oil project is far from having achieved its initial objectives. Except in circles close to the authorities, this opinion, which is also conveyed by NGOs and local newspapers, is quite widespread. In addition, for the government, someone who conducts research in oil is always regarded as being aligned with the position of NGOs, who had been very critical of the project and its management from its early beginnings. So how can a researcher work on oil without taking sides, without letting himself be taken over by one party or the other? As a native of Chad, aware of all these debates and controversies on oil, the researcher had to find a way to propose an anthropological discourse on the issue. In addition, in the cantons and villages of the oil zone, the populations seem disillusioned and hold very negative opinions on oil. How can the researcher share the lives of these populations, their difficulties and misfortunes without taking a stand? Especially since in most cases interlocutors seemed interested to know what concrete and immediate results the researcher's work would provide for them. He was to abstain from any predictions and let himself be guided by the field. Having left for the field without preconceptions, he was to formulate his questions according to the observations made, trying as much as possible to speak with all the actors concerned by the oil issue so as to take into account the controversies as they unfolded in the field. He also tried to remain as critical as possible of the data collected in order to identify a position that could be empirically defendable. To do this, the researcher constantly questioned the data, trying to triangulate them with repeated fieldwork trips, and taking up and deepening certain points. Undertaking research work on issues such as security, governance, migration civil society, citizen dynamics, often brings the same challenges to the table.

Freedom of speech and professional pressure

In addition to the challenges of research, the publication of results and speaking out in public debate on these issues is equally sensitive. The researcher who assumes such responsibility will be considered to be either an opponent to the governing regime, or an activist and, thereby subject to various pressures and obstacles in the development of their career as a teacher-researcher in the public sector of a country where private education is embryonic and of low quality. Here we reach a central point that we wish to make, namely, that the situation described above reinforces the temptation to switch definitively from research to consulting which is much better paid and morally rewarding since the researchers' skills and results are, if not locally, then most certainly internationally valued.

A rewarding journey but ...

Having had the opportunity to work on so many themes, with such diverse partners and in such diverse contexts, the few local anthropologists had the unique opportunity to discover the depth of Chad and the dynamics that have prevailed there in the last decade. Examining these dynamics has greatly contributed to strengthening all the participating researchers' anthropological understanding of the country.

Beyond that, such a journey has facilitated the development of additional skills of both CRASH's and the Department of Anthropology's researchers enabling them to interact with development and decision-making circles and thus, to contribute to knowledge production in a country where academic research has never received the attention it enjoys elsewhere. This leads us to raise the question of the usefulness of research in a country such as Chad. Do we do research because we are concerned about development? Or do we want to alert development actors to the need to understand social dynamics so as to better calibrate their actions? Should researchers accordingly improve their tools to better understand the impacts of their actions in the field? Obviously, in order to achieve changes in the spheres of development work, we need to move away from the principles of independent and disinterested fundamental research, not permanently, but at least in the near future. The challenge that has been taken up at CRASH is to take advantage of the fact that development actions create change and constantly bring new areas of interest on which to focus analysis. Like this, an enrichment occurs in both directions.

A model for combining action research and academic research

With this contribution, we underline how anthropology in Chad has, and continues to be, a non-linear journey as it is poorly institutionalised and lacks any real governmental support – consequently it is not easy to plan ahead. Such a journey requires not only great flexibility, but also a certain ability to adapt and become attuned to what is at hand. In particular, combining academic research with consultancy work brings the temptation of taking on too many tasks at the same time, and ending up doing them badly.

Dependence on external sources of funding and consultancy leads to a certain thematic nomadism, which, if not controlled, prevents further theoretical development. Unfortunately, our research so far remains tightly constrained to the factual demands exerted by external funders, particularly when it comes to moving from empirical findings to attempts at complex theorisations. Therefore, bearing in mind the dilemma of poor funding in Chadian academia and no governmental support on the one hand, and high international interest in quality surveys and qualitative expert reports on the other, we suggest a practicable model. We see that being able to specialise on a limited number of topics and themes becomes difficult unless we can work together in a larger group of well-trained scholars. As soon as that situation is achieved, it would allow different researchers to gradually identify appropriate and more specific research fields. Currently, we are trying to establish this kind of specification at CRASH, by recruiting more and more of the best students from the master's programme. We also orient students towards particular themes as soon as they have chosen their subjects for a master's thesis. Additionally, we recruit Chadian students from other fields of the social sciences who work on themes of interest to CRASH and its partners. Only by enlarging the team and developing specific interests and specialisations will it become possible to reconcile well paid action research and underfunded academic research. By making it possible for trained researchers to combine action research and academic interests following a specialized topic of research, the centre will be able to promote students from the masters to higher levels of academic achievement. A scholar who writes a thesis concerning the Internet and Communication Technology viewed in relation to development, might, for instance, later specialises on new markets and start-ups, or someone interested in public health would investigate exclusively within the medical and health related sector. Similarly, students of religious diversity, of conflict, or of gender issues, would have opportunities to attain the higher academic levels so that apart from writing reports in consultancy contracts, they would also be able to publish their findings in academic journals, and thus achieve greater international visibility.

Conclusion

We started this paper by describing our own experiences within a collaborative research project, the "Travelling Models in African Conflict Management" project. Expanding on the varying expectations of young researchers from African countries, who were keen on "making models work for development" and their African and German supervisors who were interested mainly in "academic achievements and the advance of theory", we have tried to suggest creative ways of combining these two avenues. And even though this idea is born from the lack of research funding in Chad, and the need to find it through consultancies internationally, we also aim at rendering different ways of pursuing anthropology internationally accepted and valued. This accords with Law's and Lin's (2017) demand for a more symmetrical and multi-versed anthropology, one that provincialises the current domination of its western-based themes and representatives. In line with this is Ribeiro and Escobar's (2006) call for "World Anthropologies". Such anthropologies do not start out from any particular national standpoint, but develop across a heterogenous, democratic and transnational anthropological community. Many might consider this utopian – a means to do away with purely academic standards of doing research, of publishing and of passing through the eye of the needle to reach positions of highest reputation. CRASH's model of combining what is practical with that which is of the greatest meaning and validity requires the development of a group of engaged academic researchers and teachers who study and understand the situation in their country. Ideally, such a research community would work towards enlightening not only the international community, but also Chad's own youth and its governmental bodies. This tactic would amply demonstrate the usefulness of anthropological inquiry. But what would this model mean for the western based academic, eager to collaborate with Chadian researchers – and also for those who do not (yet) know of the rich political-religious-historical-social-cultural-economic fields of knowledge to be gained in this area of the world? A starting point would be to accept the different forms of becoming an expert, be it as a well-funded academic researcher or as a well-paid consultant. It would also be important to recognise the value of working hard to establish a specific field of expertise whilst acknowledging and acting on the need to nomadically wander from theme to theme. With engaged and activist anthropology becoming more and more accepted, the practice and outcome-oriented research of consultant-academics needs to be accepted as a valid avenue of academic achievement, and everyone in academia should stop being disparaging about this route, now. If Chadian anthropology currently needs to be more "applied" than "purely academic" – and more concerned with justice then global traditions – this independence is of considerable relevance for any effort at decolonisation. After all, decolonisation is an abstract term, and only measures such as those we have described here can invest the project of decolonisation with practical and meaningful life.

Bibliography

Alwihda Info (2023): Tchad : 57.443 étudiants dans les établissements publics en 2022 et 2700 enseignants-chercheurs. https://www.alwihdainfo.com/Tchad-57-443-etudiants-dans-les-etablissements-publics-en-2022-et-2700-enseignants-chercheurs_a121058.html (last access 20 May 2024).

Apter, A. (2005): The Pan-African Nation : Oil and the Spectacle of Culture in Nigeria, Chicago/London: The University of Chicago Press. https://doi.org/10.7208/chicago/9780226023564.001.0001.

Behrends, A. (2015): Gemeinsam Forschen. Die Ethnologie als 'Travelling Model' der Forschungspraxis und Theoriebildung. (https://allegralaboratory.net/vortragstitel-gemeinsam-forschen-die-ethnologie-als-travelling-model-der-forschungspraxis-und-theoriebildung-2/)

Behrends, A./Hoinathy, R. (2017): "The Devil's Money: A Multi-Level Approach to Acceleration and Turbulence in Oil-Producing Southern Chad." In: Social Analysis 61/3, pp. 56–72. https://doi.org/10.3167/sa.2017.610304.

Behrends, A./Park, S.-J./Rottenburg, R. (2014): "Travelling Models: Introducing an Analytical Concept to Globalisation Studies." In: A. Behrends/S.-J. Park/R. Rottenburg (eds.), Travelling Models in African Conflict Management. Translating Technologies of Social Ordering, Leiden: Brill, pp. 1–40. https://doi.org/10.1163/9789004274099_002.

Behrends, A./Reyna, S. P./Schlee, G. (eds) (2011): Crude Domination: An Anthropology of Oil, Berghahn Books. https://doi.org/10.3167/9780857452559.

Chakrabarti, D. 2008 (2000): Provincializing Europe. Postcolonial Thought and Historical Difference, Princeton: Princeton University Press. https://doi.org/10.1515/9781400828654.

Coronil, F. (1997): The Magical State: Nature, Money, and Modernity in Venezuela, Chicago & London: University of Chicago Press.

Ferguson, J. (2005): "Seeing like an Oil Company. Space, Security and Global Capital in Neoliberal Africa." In: American Anthropologist 107, pp. 377–382. https://doi.org/10.1525/aa.2005.107.3.377.

Hoinathy, R. (2012): Pétrole et changement sociale. Rente pétrolière et monétisation des relations économiques et sociales dans la zone pétrolière de Doba, Paris: Karthala.

Hoinathy, R. (2022): Chad's Political Transition Lags Behind Deadline. (https://issafrica.org/iss-today/chads-political-transition-lags-behind-deadline).

Hoinathy, R./Behrends, A. (2014): "Does Rationality Travel? Translating a World Bank Model for fair Revenue Distribution in Chad." In: A. Behrends/S.-J. Park/R. Rottenburg (eds.), Travelling Models in African Conflict Management. Translating Technologies of Social Ordering, Leiden: Brill, pp. 76–91. https://doi.org/10.1163/9789004274099_004.

Hoinathy, R./Betinbaye, Y. (2023): Sudan's War Poses Grave Risks for Chad and CAR. (https://issafrica.org/iss-today/sudans-war-poses-grave-risks-for-chad-and-car).

Hoinathy, R./Janszky, B. (2017): "The Extractive Industries Transparency Initiative (EITI): The Latest Attempt at Governing the Extractive Industries in Chad." In: The Extractive Industries and Society 4/4, pp. 825–832. https://doi.org/10.1016/j.exis.2017.11.004.

Latour, B. (2004): "Why Has Critique Run out of Steam? From Matters of Fact to Matters of Concern." In: Critical Inquiry 30, pp. 225–248. https://doi.org/10.1086/421123.

Law, J./ Lin, W.-yuan (2017): "Provincializing STS: Postcoloniality, Symmetry, and Method." In: East Asian Science, Technology and Society: And International Journal 11, pp. 1–17. https://doi.org/10.1215/18752160-3823859.

Onyango-Ouma, W. (2006): "Practicing Anthropology at Home: Challenges and Ethical Dilemmas." In: Babiker, M/Mills, D./Ntarangwi, M. (eds), African Anthropologies: History, Critique and Practice (Africa in the New Millennium Series), New York: Zed Books, 250–266.

Ramoupi, N. L. l./Ntongwe, R. N. (2016): A Critique of the Discourse on Africanization of Knowledge in the Universities in Africa: Africanization of Humanities Knowledge in the Universities in Africa: The whys and how far in Cameroon and South Africa, Johannesburg: WITS Institute for Social and Economic Research (University of Witwatersrand). https://doi.org/10.1007/978-94-6300-842-6_12.

Reyna, S. P./Behrends, A. (2011): The Crazy Curse and Crude Domination: Towards an Anthropology of Oil. In Behrends, Andrea/Reyna, Stephen P./Schlee Günther (eds), Crude Domination: An Anthropology of Oil, New York: Berghahn, 3–29. https://doi.org/10.3167/9780857452559.

Ribeiro, G. L./Escobar, A. (2006): "World Anthropologies: Disciplinary Transformations within Systems of Power." In: Gustavo Lins Ribeiro/Arturo Escobar (eds.), World Anthropologies: Disciplinary Transformations within Systems of Power. Oxford: Berg, 1–27. https://doi.org/10.4324/9781003087434-1.

Rottenburg, R. (2009): Far-Fetched Facts. A Parable of Development Aid. Cambridge, MA : MIT Press. https://doi.org/10.7551/mitpress/9780262182645.001.0001.

Seli, D. (2014): (Dé)connexions identitaires hadjeray: les enjeux des technologies de la communication au Tchad, Langaa/ASC Series.

Decolonising the "Native", Insider and Outsider Categories in Anthropology

Souleymane Diallo and Karim Zafer

Introduction

In writing this chapter, we aim to contribute to the debate on decolonising anthropology in general. But, more specifically, we intend to challenge some persisting assumptions about the "native", insider, and outsider categories that still, uncritically, prevail in scholars' claims to and assessments of ethnographic authority (cf. Narayan 1993). Our argument is based on a three-step analysis. First, we outline and discuss selected examples of persistent and uncritical assumptions that still inform some scholarly assessments of ethnographic authority. Second, we will bring these assumptions into conversation with examples from our fields. In doing so, our aim is to further problematise the empirical foundations of "native", insider and outsider categories. Third, by way of conclusion, we offer some reflections on decolonising anthropological accounts of alterity.

We situate our endeavour within the scholarship on decolonising a discipline which is historically implicated in Western constructions of alterity (Bohannan 1966; Amselle 1990). On this topic, scholars have acknowledged that anthropology has offered one of many sites for the elaboration of the radically different other, defined by tradition as "simple societies", or "societies without history" which stand in contrast to Europe (Comaroff and Comaroff 1991, 1992, 1997, 2012). However, there is no doubt that anthropology has in the past decades moved away from its colonial legacy, hoping to free itself from the conceptual and political errors of colonial-era anthropology (Rouch 1978; Ferguson 2006). Accordingly, anthropologists have successfully invested substantial effort into the conceptual reorientation of the discipline (Abu-Lughod 1991).

Rouch's mid-20th century proposition of "shared anthropology" is, among others, a good illustration of the decolonising efforts of the discipline. For him, "shared anthropology", implies a radical empirical anthropology that does not privilege theory over description, or prioritize thoughts over feelings, but rather promotes a sensual anthropology in which scholars no longer act as passive observers during their efforts to describe social life from the perspective of the other (Stoller 1992). In many respects, Jean Rouch's achievements anticipated the debate known later as the reflexivity turn (Hastrup 1995; Holliday 2003; Cousin 2010).

Rouch's insights and the reflexivity turn moved toward an anthropology of relationality thus departing from an anthropology that creates the other as radically different by making this other its object of study or *raison d'être* (Amselle 1990). Abu-Lughod (1991) demonstrates that the earlier anthropological usage of the notion of culture enforced separations and, by implication, created a sense of hierarchy. To counteract this tendency, she proposes three modes of writing ethnographic accounts: discourse and practice; connections; the anthropology of the particular. Abu-Lughod's second proposition is of relevance to what follows here as she urges anthropologists to focus on historical connections that tie the people under study to the world beyond. This emphasis counterposes those accounts that focus on differences or separations (ibid.:148-149). We draw on Abu-Lughod's proposition of an anthropology of connections to challenge particular categories informed by colonial ideations of alterity that remain prevalent in scholarly claims to, and assessments of, ethnographic authority.

Language, alterity and ethnographic authority *à la fin du siècle*

We begin our discussion by reflecting on how scholars assess ethnographic authority. We aim to demonstrate how these assessments promote an anthropology of separation rather than an account of connections. We take account of connection as an effort toward decolonial anthropology (Abu-Lughod 1991). Our analysis starts with a reviewer's comments written four years ago regarding an article submission made by Diallo. The reviewer stated:

> "...I would also advise the author to include a description of research methodology. From this submission, we only know that the author spoke to and observed two households in Niamey, and conducted several con-

> versations between 2012 and 2017. For how long did the author conduct fieldwork in Niamey? How many people was the author able to observe and speak with? How well did the author speak Tamashek? Without answers to these questions, it is impossible to judge the authority of the author's claims about this community in exile."

The reviewer's comments informed the editorial decision stating:

> "I regret to inform you that the reviewers have raised serious concerns, and therefore, your paper cannot be accepted for publication ... Since two reviewers do find some merit in the paper, I would be willing to reconsider if you wish to undertake extensive revisions and resubmit, addressing the reviewers' concerns. Their recommendations include refining your argument, drawing on a wider body of literature, and explaining your methods in more detail...".

To clarify, the submission has subsequently been revised, accepted, and successfully published by the same journal. Therefore, the point of our analysis is not to fight back because the submission was rejected – quite the opposite. We found the exchange between the reviewers, the editorial board and the author productive and relevant to our argument here.

Let us begin with what we saw as implicit assumptions contained in the reviewer's comments. The reviewer raised a number of questions. For instance: how long did the author conduct fieldwork in Niamey? How many people was the author able to observe and speak with? We do not dispute these. What we want to dispute is one particular question and the concluding statement that follows it: "How well did the author speak Tamashek? Without answers to these questions, it is impossible to judge the authority of the author's claims about this community in exile". Clearly, the initial editorial decision to reject the piece prompted the reviewer to request more elaboration on the methodological section of the submission.

For us, the reviewer equates the ethnographic authority of the author with the ability to fluently speak "the language" of the people under study. This assumption is problematic as it implies that the Tuareg who participated in this research only speak Tamashek. This perspective fixes the Tuareg who participated in this research in space and time and is thus reminiscent of colonial understandings and classifications of the native people of Africa and elsewhere (see Amselle 1990; Mamdani 1996). Our point is that this portrayal does not correspond to an empirical fact. What it silences is the reality that like the Turkana, Masai, and Pokot in East Africa who speak each other's languages, the Tuareg observed in Niamey also speak several other languages. These include Bamanankan spoken in southern Mali, Houassa, Zarma, and Songhay spoken in northern Mali and Niger. So how did people acquire these languages? Some learned them through social transactions with their neighbours. Many went to school at an early age and learned to speak French. As a result, during fieldwork, the Tuareg who participated in interviews and focus group discussions used the languages mentioned above as well as Tamashek. As shown below, we consider that conducting fieldwork in several languages allows a dynamic ethnographic account of connection (Amselle 1990; Abu-Lughod 1991; Scheele 2012) freed from colonial classifications that tend to freeze the people under study in space and time.

Reflections on language and connections in our fields

The first example presented here draws on Diallo's fieldwork among the Tuareg from northern Mali in Niger. The Tuareg are historically a pastoral nomadic group living in the southern fringe of the Sahara and in the northern Sahel. Geographically, this area includes southern Algeria, northern regions of Mali, Niger, and Burkina Faso. The Tuareg living in Mali are homogenous. They are politically and socially organised around various rival federations that refer to themselves as Kel Adagh, Ouillimiden Kel Ataram, and Kel Tinguériguif and Kel Antsar in the regions of Kidal, Gao, and Timbuktu, respectively. The term "federation" (in Tamashek, *ettebel*) refers to a set of clans that form a political unit under the leadership of one dominant clan (Lecocq 2010: 13). The different clans that make up one *ettebel* stand in hierarchical relationship to each other. The leader of the ettebel is known as the *amenokal*, which means "the owner of the land" and he is also invested with [political?] power. Several federations rose and fell during the 18th and 19th centuries (Grémont 2010). These groups of Tuareg speak different dialects of the Tamashek language. They define themselves as the Kel Tamashek, those who speak Tamashek (Lecocq 2010).

As we have said earlier, although Tamashek is the language primarily spoken by the Tuareg, many of them use other vernacular languages spoken by neighbouring groups, such as Fulani and Songhay, and increasingly French and English are used as a result of schooling and transnational migrations. Also, like many other pastoral communities in Africa, the Tuareg's former nomadic and semi-nomadic lifestyle has been irreversibly altered in the course of the 20th century. These alterations resulted from the repercussions of recurrent periods of conflict and drought, while famines that occurred during the last half-century induced exile, along with the persecution or simple neglect of northern populations by the Malian central state (Lecocq 2010; Klute 2013). These complex dynamics jointly prompted the relocations of many Tuareg in the West African diaspora where they learned to speak several other languages (Diallo 2018).

During the research conducted by Diallo between 2012 and 2024 in Niger, the methods consisted of participant observation, narrative interviews, and focus group discussions. Narrative interviews and focus group discussions were conducted in Tamashek, French, and occasionally in Bamanankan with Tuareg who had served in the Malian army or worked as school teachers in Mali prior to their migration to Niamey. These Tuareg research participants were eager to answer questions in Bamanankan, Diallo's mother tongue. The reason for that was obviously not primarily the researcher's inability to speak in Tamashek. Rather, it was to show that they could also speak other languages and had been interacting with non-Tuareg groups. For instance, a female informant, named here Fatoumata, several times insisted

> "I am so glad to talk to you in Bamanankan. It has been my main language for several years. I grew up with my sister who is married to a civil servant from the south. He is a Bambara. We lived in Kidal, Menaka, Timbouctou, Kayes, and Bamako. I spent 15 years with them. I'm glad I could pick up this language now again. Just ask me your questions, I would prefer answering them in Bamanankan."

Fatoumata spoke not only Bamanankan and Tamashek, but also Songhay. For us, the question that arose from this case and several others is as follows: why would speaking to Fatoumata in Bamanankan rather than Tamashek undermine [or challenge] the ethnographic authority of the researcher? Our answer to this question is that speaking these other languages does not make Fatoumata less Tuareg than others who don't have this ability – quite the contrary. Fatoumata's case substantiates our point that her mother tongue (Tamashek) is not the only communicative idiom for her in particular or for the Tuareg in general. Ultimately, this prompts us to re-conceptualize native/informant identity and positionality in a new and useful way.

Our second example draws on Zafer's work and his research participants' abilities to understand each other's Arabic despite dialect differences.

Zafer, originally from Egypt, conducted fieldwork in the Federal State of North Rhine-Westphalia in Germany with unaccompanied minors and youth refugees from Syria and Iraq. We believe this example offers another fertile area for the elaboration of an anthropology of connections rather than the differences that freeze social actors in time and space.

In linguistic terms, the "Arabic language is a collection of spoken dialects with important phonological, morphological, lexical, and syntactic differences, along with a standard written language, Modern Standard Arabic (MSA)" (Chiang, Diab et al. 2006: 369). Although the Arabic language is the official language for all Arabic countries, each country and even each region in the same country has a different dialect. Those dialects are what people use in their daily interactions, and not "Modern Standard Arabic". The standard written language, which is mainly used in official discourses and texts, is the same throughout the Arab world. Chiang, Diab et al. (2006) add that MSA is based on Classical Arabic and is not a native language of any Arabic-speaking people, therefore children do not learn it from their parents, but in school. They add that most native speakers of Arabic are unable to produce spontaneous MSA. Chiang, Diab et al. also add that dialects vary not only along a geographical continuum, but also in accordance with other sociolinguistic variables such as the urban/rural/Bedouin dimensions (ibid.).

From Zafer's personal experience and encounters in the field, we would say that some dialects may resemble others, while others might be totally incomprehensible to people outside the region or the country. For instance, dialects in the Maghreb (Morocco, Tunisia, Algeria) may, to some extent, resemble each other for someone not from the Maghreb, and are very difficult to follow for people outside this region. By the same token, dialects from Syria, Lebanon, Palestine and Jordan are, to a large extent, easily understandable within the different populations of the Levant Region; the same might also apply to dialects in the Gulf States. The Egyptian dialect, the one Zafer speaks, is the most popular and understandable one, mainly because of the popularity of many Egyptian singers, films, and theatre productions in the Arab world.[1] Additionally, most Egyptians are able to easily understand Levantine Arabic and to a lesser extent Gulf States dialectics, yet they experience difficulties in understanding dialects from the Maghreb. Since Zafer's research participants are

1 It should be also noted that people across the MENA region tend to claim that their dialect is the most comprehensible and purer one. Purer here refers to proximity to classical Arabic, *al-fusha*, which MSA also refers to.

mainly from Syria, Palestine, and Iraq, he is able to understand and follow their dialects, just as they can understand his Egyptian/Cairene dialect. Furthermore, while talking with him, they tend to use some popular Egyptian expressions due to the popularity of Egyptian singers and actors they are fans of. To return to the point of the "insider researcher"– that is those who share a linguistic heritage or knowledge with the group being studied – this designation cannot be fully applied in Zafer's case. The heterogeneity of Arabic dialects controverts the uncritical assumption of the "native" which assumes that all people from Arab countries speak the same language and dialect. This perspective not only ignores the existence of internal variations within the Arab countries, but also fails to acknowledge and interpret the capacity to understand all dialects and thus overlooks the potential of social actors from these countries to transcend their differences. Indeed, there is a great deal of stratification in speaking Arabic and claims of being "native" are always relational and prone to contestation. We argue that the acknowledgement of this potential makes a positive step towards an anthropology of connections.

The positionalities of insider and outsider in our field settings

In the previous section of the chapter, we focused on language in order to examine the shifting positionality of the research participants and to advocate for an anthropology of connection. In what follows, our analysis will focus on the researcher's positionality in relation to the ethnographic account of connection. Our analysis is structured around the notions of "insider" and "outsider" as both terms are often evoked to depict the researcher's positionality during fieldwork. We discuss the ways in which these terms have informed anthropological representations and the ethnographic authority of the author (Spradley 1980). Holliday describes the concept of "insider" to define the position of a researcher doing fieldwork in his own culture. The concept of "outsider" refers to a researcher who is on unknown terrain (Holliday 2003). These notions are mostly deployed to make sense of the researchers' positionalities in the field. Whereas African researchers doing fieldwork in their own local settings are considered to be insiders, their European counterparts are conceived of as outsiders. By the same token, Nowicka and Ryan write: "insider researchers share a cultural, linguistical, ethnic, national and religious heritage with their participants" (Nowicka and Ryan 2015: 2). They stress, however, that it should not be forgotten that ethnic and national belonging and gender are multi-layered, culturally constructed concepts (ibid.). In our view, this understanding promotes ethnographic accounts which are predicated on separation rather than connection.

For instance, in an influential contribution, Diawara has argued that while African researchers working in their own national contexts face fewer problems with language and living conditions, they might encounter more significant difficulties if compared to their counterpart Western scholars doing fieldwork in Africa (Diawara 1985). Speaking about his research (collection of oral history) among his own people in southern Mali, he notes that his being a descendent of the local nobility and from a politically dominant clan raised suspicion among the people of servile background whose oral accounts and perspectives he sought to investigate. Moreover, some scholars are in favour of abandoning the dichotomy of insider-outsider research as this scheme prioritizes one particular kind of difference – most commonly the ethnic or national – over other categories of difference.

We argue here, however, that insider and outsider positionality within fieldwork is determined by the quality of relationships between research participants and researchers rather than by the national origins of the latter. We contend that focusing on the quality of these relationships helps to promote an ethnographic account of connections that challenges an over-emphasis of the importance of the researcher's country of origin.

Insider or outsider? Reflections on positionality and connection in the field in West Africa

The following statement, made by a research participant in Diallo's research, illustrates the complexity surrounding the researcher's own positionality among the Tuareg in Niger.

> "Before you, we saw Bajan[2] and other Tuareg with white people [read: Europeans] doing research in Menaka. As the Tuareg are clever, they never gave those researchers any chance to talk to us. Because they know that

2 Bajan is used to refer to Bajan Ag Hamatou, the political leader of the dominant clans in the area of Menaka. He is currently the deputy of Menaka at the national parliament of Mali.

> many things they are telling them are untrue. For this reason, one precaution they took was to follow these Europeans closely in order to control everything they were doing. They slaughtered goats, sheep, even camels for them, all that was done in order to keep their research agendas busy and under their close watch. Another thing is that the Tuareg followed these whites for the material benefits some of these researchers might have offered them. In the end, these researchers did not even respect us. But your case is different. You respect us because you came to spend time with us although the red Tuareg warned you not to come. Also, they see you just like a poor Malian student who cannot offer them any material benefit because as an African you need that money for your own parents, too. For this very reason, you are not attractive to them like European researchers would be" (Conversation with Assaley on 23/12/2012 in Abala).

This statement derives from a conversation with a male research participant in his early sixties. We call him Assaley. In this account, the research participant pointed to how the people under study situated themselves vis-à-vis other groups through their interactions with the researcher in a way that challenged the conventional prestige hierarchy. This conventional prestige hierarchy divides Tuareg society into three main groups. At the top level of society stand the free noble warriors. This is followed by other free-born people, known as the vassal groups. The unfree-born former slaves, who are known as the Bellah in the areas of Timbuktu and Gao, Iklan, in Kidal, occupy the bottom level of the social structure. The research participant reduced the whole notion of being Tuareg to the free-born Tuareg who, in his view, had been controlling the research done by European researchers (see Bouman 2010; de Sardan 1976).

Assaley's account assumed that Diallo being a Malian and a poor African student made it possible for him to move between different social segments without encountering major obstacles. Assaley's understanding of Diallo implied that he thought that the researcher did not have adequate material means to be able to distribute resources to the free-born white Tuareg. His argument also illustrates how the Bellah-Iklan speakers perceived themselves not only as having been marginal to political processes in northern Mali since the colonial era, but also peripheral to the generative dynamics of academic knowledge. This argument allows us to discuss Diawara's insight outlined above (Diawara 1985). As mentioned earlier, Diawara has shown that he faced particular methodological challenges when attempting to talk to people of a servile condition. He explained that these challenges were due to his noble background in his own society (ibid.). What Assaley tried to show here, which we would argue adds a new dimension to Diawara's discussion, is that [contemporary] non-African scholars may also face exactly the same dilemma that their African counterpart, Diawara, experienced during his research [four decades ago]. Assaley clearly associated the European researchers working in the area of Menaka with the politically dominant social segment of the free-born white Tuareg. To him, the European researchers were prohibited from speaking to the Bellah-Iklan because doing so could compromise the researcher's relations with the free white Tuareg who hosted them. Therefore, in Assaley's eyes, those researchers no longer occupied a neutral position if local [hierarchical] tensions were taken into account. This illustrates that as European researchers enter African societies through social contacts that consequently absorb them, they become part of the local struggles for power and legitimacy. This is precisely the feature that Diawara's account overlooks, and in so doing it fixes both non-African researchers and African societies in a rigid and static manner.

Concerning Diallo's case, although the researcher is a Malian, he is not originally from either the free white Tuareg group or the Bellah-Iklan, so in this regard, his status differs from that of Diawara who originally belonged to one of the groups on which he was doing research. In contrast to Diawara's case, our example from Niger suggests that the researcher's status becomes closer to that of non-African researchers who are not initially members of any given local groups. We would also argue that in the same way as that in which Assaley associated European researchers with the politically dominant free-born Tuareg groups in Menaka, he tended to associate the African researcher with his own social group in Niger.

This was evident in how the Bellah-Iklan research participants often overtly insulted the few free-born families living in Abala when these people were passing by and interacting with Diallo. As a result, the free white Tuareg kept their distance and did not want to interact beyond the standard exchange of greetings, especially when there were Bellah-Iklan men around. It was because of the apparent tension, which informed the free-born Tuareg and Bellah-Iklan interactions in the refugee camp that it was not possible to follow the two groups in one location, Abala. Indeed, attempting to simultaneously integrate the few free-born families as research participants in the refugee camp would have resulted in obvious tensions that could have caused problems with NGO representatives. As such, the research could have further contributed to transforming the social setting of exile into an open battleground. This was obviously not the goal of the stay there in Abala. The male Bellah-Iklan's attempts to control the interactions with the free-born Tuareg indicate the challenges faced in the field. Though these challenges might have some limiting implications for the present study, the focus on a group of free-born Tuareg in Niamey enabled the inclusion of the latter's perspective in the research.

This example from Niger displays how the relationship between the research participants and the researcher is determined by a complex constellation of processes that challenges a sharp distinction between insider and outsider positionality. It shows that a researcher may simultaneously be an outsider and an insider depending on the ways in which the research participants view him or her. This complexity is reminiscent of the argument put forward by Crawford and Hastrup, namely that the dichotomy of researcher versus research participant may lose its relevance during field research as a result of social processes that undermine a researcher's attempt to maintain his/her distinctive status (Crawford 1992: 48; Hastrup 1995). Hastrup has called this "The process of becoming the other" (Hastrup 1995: 19). For her, "becoming" here is just a metaphor to point to the kind of participation that can never be completed, and which is not an immediate consequence of physical presence. It does not attempt to argue that the anthropologist becomes identical with the others among whom he or she is doing research. For example, Hastrup herself recognizes that she did not literally become an Icelandic shepherdess, even though she participated in sheep farming and experienced the extraordinariness of shepherdesses' work in misty mountains. To her, this implies that one is not completely absorbed into the other world, but neither is one any longer the same. The relevance of Hastrup and Crawford's discussion here is that the researcher does become part of the community among whom he/she is doing research, but this is not because he or she changes. Rather, the perceptions of the people among whom the researcher is doing research can change during fieldwork in such ways that they begin to relate to him/her in familiar terms.

In our view, the discussion of Diallo's positionality prompts two strands of reflection on the connections between the researcher and the people under study. First, the way in which the Bellah-Iklan research participants perceived and ascribed a certain status to Diallo suggests that the researcher becomes an integral part of, and an active agent within, the local social universe. Second, the Bellah-Iklan interlocutors ascribed a local status to the European researchers. This invites us to think about an ethnographic account of connection in a particular way, one which involves carefully determining the nature of the relationship between the researcher and the people under study.

Insider and/or outsider? Zafer's reflections on positionality in the field, Germany

"You are an insider researcher then". This is how some anthropologist colleagues described Zafer's researcher-researched relationship despite his contestations. In fact, we see the notions of insider and outsider not as a dichotomy, but more as a continuum in which the researcher may (un)consciously move between its poles. In the following, we explore some circumstances where Zafer considered himself an insider and others where he was seen as an outsider by the research participants.

The most obvious aspect which led to Zafer being ascribed the status of insider researcher is the so-called "Arab culture identity" that he shares with the research participants. That is, he is originally from Egypt and has been conducting research with unaccompanied minors and youth refugees from Syria, Iraq and Palestine living in Germany. We will not discuss here what "Arabic culture" is or if it is correct to claim that people from Arab countries share a common culture, but perhaps it is worth noting, for instance, in recent anthropological studies, that using the term Arab-majority societies is more politically correct than Arab World since the former avoids associations of insularity and homogeneity (Deeb and Winegar 2015). Nevertheless, we argue that the term Arab-majority societies is not ubiquitously accepted since a number of Arab social scientists refuse to adopt this "politically correct" term as they believe that it is yet another Western attempt to strip the region of its identity. Indeed, one should bear in mind that the term Arab still has a meaningful social and political construction in such societies, one that affects social life for Arabs as well as for ethnic or linguistic minorities (ibid.).

With regard to the first aspect, which concerns sharing "Arab culture identity" with persons from Arab-majority societies, we believe that ideas of Pan Arabism[3] play an important role in our encounters and interactions

3 Lawson (2014) writes that Pan-Arabism refers to two different political and historical circumstances. The first, referring to Arab nationalism, emerged at the end of the 19th century among Arabic speakers inside the Ottoman Empire and asserted that "Arab" across the empire shared characteristics that entitled them to think of themselves as

with each other (see Choueiri 2007; Lawson 2014). Pan Arabism is an ideology that has shaped the perceptions of many people in Arab-majority societies, through political and public discourses, and education. For instance, in many Arab-majority societies, this topic was and probably still is an obligatory class at public schools. Arabs were raised and educated to perceive their relations to other Arab countries and societies as brotherhood relations. They were taught that prosperity and development in this region could, or should, only be achieved when Arab countries unite against imperial Western domination and conspiracies. It is true, that quite recently and even before the Arab Spring and the disappointment caused by its failure in some countries, as in the case of Egypt, strong voices from many secular intellectuals called for embracing ideas of "Egyptian Nationalism" and a refusal to identify with the "Arab culture identity". Their calls have found many supporters, especially from the middle and upper classes. Yet, according to Zafer's experience, the brotherhood between Arabs is still inscribed and dominant in the perceptions many Arab people have of each other. This sentiment of brotherhood, derived from the Pan Arabism ideology, we would argue, is a main characteristic of the so-called "Arab culture identity".

The second aspect of being an insider researcher involves being raised in and belonging to an "Arab family". Of course, we are not claiming that there is only one model of the "Arab family": we are treating this analytical category as an ideal type in Weber's sense (Weber 1969 [1949]) for the sake of comparison, especially with regard to Western family models. Being part of and belonging to an "Arab family" allowed Zafer to have a deeper understanding of the characteristics and dynamics of his research participants' family relations, whether family members were living together or maintaining transnational relations (see Joseph 1999, 2005).

Nevertheless, Zafer's gender, an important component of his multi-layered identity, made him, in some instances, an outsider researcher. Like him, all his key research participants were males. Being a male Arab researcher/brother meant that Zafer could not recruit young females as research participants. This was due to the methodology used during the fieldwork which involved being an elder brother to the research participants and "deep hanging out" with them

a single community. The second period refers to the political discourse of Arab governments in the mid-20th century. Pan-Arabism during that time presumed that the Arab world has common strategic and economic interests that transcend the interests of individual states.

(Clifford 1996; Geertz 1998). Being too close or trying to build a friendship between an Arab male and a young Arab female is not an easy task and could be problematic for both sides. In this regard, Zafer recalls two incidents which show that even having an informal talk with an Arab female in the social network of some research participants was not appreciated by all parties. The first incident is related to the romantic relationships of two key research participants, Jaber and Magdi. Jaber was in a relationship with a Syrian woman and Magdi with a Dutch one, and both decided to sign a marriage contract with the support of a Muslim sheikh. Jaber refused Zafer permission to meet and talk to his wife since both of them are Arabs. By contrast, Magdi did not oppose Zafer's wish to speak to his Dutch wife about their relationship. If Magdi had an Arab wife as Jaber did, Zafer claims that Magdi would also have refused to allow Zafer to speak with his wife. In the second incident Zafer was invited to the house of a key research participant, Mohamed, to meet his newly arrived family in Germany. Upon Zafer's arrival, the mother greeted him, but during the entire afternoon, she did not join them at all, neither for lunch nor afterwards in the living room. Hence, he did not have the opportunity to speak to her.

The third aspect with regard to Zafer's ascribed position in the field as an insider researcher is the migratory trajectory. The point here is not to claim that the research participants and Zafer shared the same migratory experience. It is by no means fair or correct to compare the participants' forced migration experience, their losses and their traumatic experiences to Zafer's migratory experience as a 'voluntary' migrant in Germany. Nevertheless, we do not deny that making a clear-cut distinction between voluntary and forced migration is a difficult task. Yet, especially in Zafer's case, important differences in migration trajectories, legal status and transnational relations between the research participants and himself are evident. Although it is possible that they shared some challenges due to their new life in Germany, these still do not mean that Zafer's position can be described as that of an insider researcher within the forced migrants'/refugees' community in Germany. It is important to note that in migration research, the so-called migrant "community" is usually defined ethnically (Ryan 2015). This was the case until the debate about going beyond the ethnic lens was initiated (Schiller, Çaglar and Guldbrandsen 2006). Accordingly, migrants should not be seen as simply insider members within a clearly defined ethnic community because far from being united and cohesive, so-called communities of migrants are divided by social fissures of class, generation and gender (Ryan 2015).

To sum up, from our encounters in the research fields, we would agree with Ryan's (2015) suggestion to move away from fixed notions of insiders and outsiders and focus instead on dynamic positionalities and relationality. We would not, however, support the call to completely abandon these notions, as suggested by Nowicka and Ryan (Nowicka and Ryan 2015; Ryan 2015) since being an insider has shaped and influenced the relationship with the research participants, the methodology and the research itself. We see the notions of insider and outsider not as a dichotomy, but more as a continuum in which the researcher may (un)consciously move between its poles. In this regard, we agree with Uddin, who argues that the position of an ethnographer must be in between the dichotomy of "we" and "they" or beyond (Uddin 2011). Or, as Merton wrote, "...we are all, of course, both Insiders and Outsiders" (Merton 1972: 22). This implies that the researchers are neither insiders nor outsiders: they are simultaneously insiders and outsiders. The particular connectivity that underlies this shift in positionality informs our understanding of an ethnographic account of connection in the sense proposed by Abu-Lughod (1991).

Conclusion: An anthropology of post "native", insider and outsider categories

In this chapter, we have attempted to explore the implications of the "native", insider and outsider category and proposed a future way forward. First, we sought to challenge the role attributed to speaking local languages in the constructions of ethnographic authority, and the usage of "native" and insider categories in anthropological accounts. Second, we argue that insider and outsider positionality in fieldwork is determined by the quality of relationships between informants and researchers rather than by the country of origin of the latter. By bringing these two aspects into one conceptual framework, we seek to decolonise the three categories of "native", insider and outsider, which are commonly used in anthropological discourses. Empirically, our claims draw on material collected and fieldwork conducted among Arab refugees in Germany and Tuareg refugees from Mali in Niger.

The chapter extends the concept of the anthropology of connection in two interrelated respects. One deals with the conceptual level. In this regard, we have shown that such an endeavour seeks to break away from holistic models that offer fixed views of participants and/or deflect researchers' attention from the individual variations present within such categories as "natives" or insiders and outsiders. The other is methodological. As an appropriate methodological approach for the anthropology of the particular and connection, we have argued for proper consideration of the relation between the researcher and the research participants. In the light of these discussions, we contend that insider and outsider positionality in fieldwork is determined more by the quality of relationships between informants and researchers than by the national origins of the latter. As a result, scholars should adopt a dynamic perspective on the people under study, a perspective that does not freeze them in time and space.

We believe that, within the decolonial project, the insider/outsider dichotomy does not and should not persist any longer, especially when this dichotomy is only based on the ethnic and cultural identity of the researcher vis-a-vis the research participants and ignores their multi-layered identities. It has unfortunately become taken for granted that an anthropologist from the Global South is an insider researcher when doing fieldwork back home or among diasporic communities from his/her home region. Students and junior researchers from the Global South, including ourselves, would not really oppose such an assumption or ascription, a response that definitely affects our options regarding the research groups we choose and the research questions we formulate. It seems that until now accepting and following, whether consciously or unconsciously, this association and ascription has been the easiest and maybe the only possible way to enter the academic field in the Global North and to prove ourselves as trustworthy anthropologists and scholars. We fully agree with Tuhiwai Smith's (1999) claim that decolonisation is an unfinished business. Yet, after many decades and many debates and reflections on the notion of being "native" and insider/outsider categories, we are still at the same point. Therefore, our contribution is a reminder of the necessity of deconstructing the notion of "native" and insider/outsider categories. It is a first step, but one where we still find ourselves stuck. The second step in the decolonisation project is to further promote anthropological research on Western communities moving beyond the traditional empirical focus on non-Western societies.

Bibliography

Abu-Lughod, L. (1991): "Writing against Culture." In: R. G. Fox (ed.), Recapturing Anthropology working in the present, Santa Fe: School of American Research Press, pp. 137–162.
Amselle, J.L. [1990] (1998): Mestizo Logics: Anthropology of Identity in Africa and Elsewhere, Stanford: California University Press.
Bohannan, L. (1966): Natural History Magazine (http://www.naturalhistorymag.com).
Bouman, A. (2010): "The Price of Marriage: Shifting Boundaries, Compromised Agency and the Effects of Globalization on Iklan Marriages." In: Anja Fischer/ Ines Kohl (eds.), Tuareg Society within a Globalized World, I.BTauris & Co Ltd. https://doi.org/10.5040/9780755610914.ch-008
Chiang, D./Diab, M./Habash, N./Rambow, O./Shareef, S. (2006): Parsing Arabic Dialects, Paper presented at the 11th Conference of the European Chapter of the Association for Computational Linguistics.
Choueiri, Y. M. (2007): Nationalisms in the Middle East: The Case of Pan-Arabism. In Y. M. Choueiri (ed.), A Companion to the History of the Middle East, Blackwell Publishing Ltf., pp. 291–312. https://doi.org/10.1002/9780470996423.ch16
Clifford, J. (1996): Anthropology and/as Travel. *Etnofoor*, 9/2, pp. 5–15.
Comaroff, J./Comaroff, J. (1991): Of Revelation and Revolution, Vol. 1: Christianity, Colonialism, and Consciousness in South Africa, Chicago: University of Chicago Press. https://doi.org/10.7208/chicago/9780226114477.001.0001
Comaroff, J./Comaroff, J. (1992): Ethnography and the Historical Imagination, Boulder: Westview Press.
Comaroff, J./Comaroff, J. (1997): Of Revelation and Revolution, Vol. 2: The Dialectics of Modernity on a South African Frontier, Chicago: University of Chicago Press. https://doi.org/10.7208/chicago/9780226114675.001.0001
Comaroff, J./Comaroff, J. (2012): Theory from the South, or how Euro America is Evolving toward Africa, London: Boulder. https://doi.org/10.1080/00664677.2012.694169
Cousin, G. (2010): "Positioning Positionality: the Reflexive Turn." In *New approaches to qualitative research*. Routledge, pp. 25–34.
Crawford, P. I. (1992): "Film as Discourse: The Invention of Anthropological Realities." In: P. I. Crawford/D. Turton (eds.), Film as Ethnography, Manchester: Manchester University Press, pp. 66–81.

Deeb, L./Winegar, J. (2012): Anthropologies of Arab-Majority Societies. Annual Review of Anthropology, 41, pp. 537–558. https://doi.org/10.1146/annurev-anthro-092611-145947

Diallo. S. (2018): "The Truth about the Desert": Exile, Memory, and the Making of Communities among Malian Tuareg Refugees in Niger, London: Modern Academic Publishing-Ubiquity Press. https://doi.org/10.16994/bai

Diawara, M. (1985): "Les recherches en histoire orale menées par un autochtone, ou l'inconvénient d'être du cru." In: Cahiers d'études africaines, 25/97, pp. 5–19. https://doi.org/10.3406/cea.1985.2183

Ferguson, J. (2006): Global Shadows. Africa in the Neoliberal World Order, Durham: Duke University Press. https://doi.org/10.1515/9780822387640

Geertz, C. (1998): Deep Hanging out. *The New York review of books, 45/16*, pp. 69–72.

Grémont, C. (2010): Les Touaregs Iwellemmedan (1647–1896). Un ensemble politique de la boucle du Niger, Paris: Karthala.

Hastrup, K. (1995): A Passage to Anthropology: Between Experience and Theory, London: Routledge.

Holliday, Adrian (2003): "Writing about Relations." In: Adrian Holliday (ed.), Doing and Writing Qualitative Research, London: SAGE.

Joseph, S. (2005): The Kin Contract and Citizenship in the Middle East. In M. Friedman (Ed.), *Women and Citizenship*, Oxford Scholarship Online. https://doi.org/10.1093/0195175344.003.0009

Joseph, S. (Ed.) (1999): Intimate Selving in Arab Families: Gender, Self, and Identity. Syracuse, NY: Syracuse University Press.

Klute, G. (2013): Tuareg-Aufstand in der Wüste: ein Beitrag zur Anthropologie der Gewalt und des Krieges (Siegener Beiträge zur Soziologie, 12), Köln: Köppe.

Lawson, F. H. (2014): "Pan-Arabism." In: M. T. Gibbons (ed.), The Encyclopedia of Political Thought, Malden, MA: Wiley-Blackwell, pp. 2642–2644. https://doi.org/10.1002/9781118474396.wbept0744

Lecocq, B. S. (2010): Disputed Desert: Decolonisation, Competing Nationalisms and Tuareg Rebellions in Mali, Leiden: Brill. https://doi.org/10.1163/ej.9789004139831.i-433

Mamdani, M. (1996): Citizen and Subject: Contemporary Africa and the Legacy of Late Colonialism, Princeton University Press.

Merton, R. K. (1972): Insiders and Outsiders: A Chapter in the Sociology of Knowledge. *American Journal of Sociology, 78/1*, pp. 9–47. https://doi.org/10.1086/225294

Narayan, K. (1993): "How Native Is a "Native" Anthropologist?" In: American Anthropologist, 95, pp. 671–686. https://doi.org/10.1525/aa.1993.95.3.02a00070

Nowicka, M./Ryan, L. (2015): "Beyond Insiders and Outsiders in Migration Research: Rejecting A Priori Commonalities. Introduction to the FQS Thematic Section on "Researcher, Migrant, Woman: Methodological Implications of Multiple Positionalities in Migration Studies." In: Forum Qualitative Sozialforschung, 16/2.

Rouch, J. (1978): On the Vicissitudes of the Self: the Possessed Dancer, the Magician, the Sorcerer, the Filmmaker, and the Ethnographer. *Studies in the Anthropology of visual Communication*, 5/1, pp. 2–8. https://doi.org/10.1525/var.1978.5.1.2

Scheele, J. (2012): Smugglers and Saints of the Sahara. Regional Connectivity in the Twentieth Century, New York: Cambridge University. https://doi.org/10.1017/CBO9781139135412

Schiller, N. G./Çaglar, A./Guldbrandsen, T. C. (2006): "Beyond the Ethnic Lens: Locality, Globality, and Born-again Incorporation." In: American Ethnologist, 33/4, pp. 612–633. https://doi.org/10.1525/ae.2006.33.4.612

Smith, L. T. (1999): Decolonizing Methodologies: Research and Indigenous Peoples, London: Zed Books Limited.

Spradley, J. P. (1980): Participant Observation, New York: Holt, Rinehart and Winston.

Stoller, P. (1992): Cinematic Griot. The Ethnography of Jean Rouch, Chicago: University of Chicago Press.

Uddin, N. (2011): "Decolonising Ethnography in the Field: an Anthropological Account." International Journal of Social Research Methodology, 14/6, pp. 455–467. https://doi.org/10.1080/13645579.2011.611382

Weber, M. (1969 [1949]): The Methodology of the Social Sciences. Translated and edited by Edward A. Shils and Henry A. Finch, New York: The Free Press.

Authors' Biographies

Abimbola O. Adesoji works as a teacher and researcher in the department of History at the Obafemi Awolowo University in Ile-Ife, Nigeria. The Historian did his doctorate at the same University and was a Team Member of the Research on Abisinwin: Interrogating the Trajectory and Treatment of Post-Partum Disorder Syndrome among the Yoruba of Southwestern Nigeria in Colonial and Postcolonial Period at Bowen University, Iwo, Nigeria in 2022. From 2009 to 2010, and subsequently in 2013, 2017 and 2021, he was Georg Forster Fellow of the Alexander von Humboldt Foundation at the Institute of Ethnology, Johann Wolfgang Goethe University, Frankfurt. His research focuses on Traditional and Contemporary Yoruba History and Socio-political History of Nigeria. – aadesoji2@yahoo.com

Heike Becker is a social and cultural anthropologist, who obtained a PhD from the University of Bremen in Germany. Since 2001 she has been teaching at the University of the Western Cape in South Africa. Since coming to UWC, she has been committed to developing new perspectives in the substance and pedagogies of anthropologies from the South. Becker has written on Southern African anthropologies, including, "Fieldwork in shared spaces: positionality, power and ethics of citizen anthropologists in southern Africa" (2005, with Emile Boonzaier and Joy Owen), "De-provincialising South African anthropology: on relevance, public culture & the citizen anthropologist" (2007), "The Past(s) and Future(s) of South African Anthropology" (2022), "Decolonise": The fall of a colonial statue, student protests and trajectories of anthropology in South Africa" (2024); as well as an entry on South African anthropology in the Wiley encyclopaedia of Anthropology (2018; with Andrew D. Spiegel). Becker has also published widely on culture and postcolonial politics, memory politics, popular culture, gender, and social and decolonial movements, with a focus on Namibia and South Africa, and more recently also on Germany. Becker currently coordinates an international collaborative project *Decolonising the Public Space in Germany and its Former African Colonies: Memory, Civil Society and the Arts.* – hbecker@uwc.ac.za

Michael Bollig works as a professor of social and cultural anthropology at the University of Cologne. He did his doctorate at the University of Tübingen on violence and conflict management in northern Kenya. He started at the University of Cologne in 2000 and established a focus on environmental challenges and socio-cultural dynamics in rural Africa. He has been speaker (and co-speaker) of the collaborative research centre (CRC) Arid Climate, Adaptation and Cultural Innovation in Africa (2000–2007) and the CRC Future Rural Africa (2018-ongoing). Between 2011 and 2015 he was vice rector for international affairs at the University of Cologne. – michael.bollig@uni-koeln.de

Daniele Cantini is a social anthropologist with years of research experience in different Arab countries, currently fellow at the Global Development Network, and affiliated researcher at the Leibniz Zentrum Moderner Orient in Berlin. He earned his PhD in social anthropology at the University of Modena (Italy) in 2006, with a thesis on the Jordanian university system and its students. In 2017 he obtained the Italian national habilitation to become associate professor in Social and Cultural Anthropology, and in 2019 the same qualification for Middle East Studies. He has held research positions, as affiliated, stipend-holder or institute member, in Cairo (Egypt), Modena, Halle (Germany), Beirut (Lebanon), and Zurich (Switzerland), and has taught courses on the anthropology of the Middle East, social and political anthropology, history and cultures of the Middle East, among others, at the University of Modena, Milano-Bicocca and Halle, at the BA and MA level. – daniele.cantini@gmail.com

Souleymane Diallo is a lecturer in Social and Cultural at the University of Münster, Germany. HIs publications include the monograph *"The Truth about the Desert": Exile, Memory, and the Making of Communities among Malian Tuareg Refugees in Niger* (Modern Academic Publishing (MAP), 2018), "The Truth is that we are not Tuareg: Exile, Memory, and the Making of Bellah-Iklan Community in Niger",dans B.S. Lecocq et A.Niang(dir); *Identités sahéliennes en temps de crise, Histoires, enjeux et perspectives,* (Münster, Lit Verlag 2019), Das „Festival du Desert" als kulturelle Praktik der Politischen Mediation in Mali," In: Zivilgesellschaft in Afrika, edited by Annette Zimmer, Markus Rohde, und

Ansgar Klein, *Forschungsjournal Soziale Bewegungen*, 37(2):196-208(2024), and several articles on changing social identities among Tuareg refugees in Niger. – s.diallo@uni-muenster.de

Hans Peter Hahn, Professor of Anthropology at the Goethe University Frankfurt/Main, works on material culture, museums, consumption, migration and mobility with a regional focus on West Africa (Burkina Faso, Ghana, Togo). His current research addresses the concept of "value in things" and consequences of the digitisation of collections. In the 2022/23 academic year, he was a visiting scholar at the University of Ghana in Accra (Legon) and worked with his Ghanaian partner on "What is Restitution?". – hans.hahn@em.uni-frankfurt.de

Remadji Hoinathy joined the Institute for Security Studies (ISS) in 2019 as a senior researcher (based in N'Djaména, Chad) in the Lake Chad Basin Programme and then as a senior Researcher for Central Africa and the Great Lakes from January 2023. Previously, he headed the Centre de Recherches en Anthropologie et Sciences Humaines (CRASH) in N'Djaména, Chad. A former fellow of the Max Planck Institute for Social Anthropology (Halle-Saale, Germany), Hoinathy holds a PhD in anthropology from the Martin Luther University of Halle-Wittenber (Germany). He was an associate lecturer in the Department of Anthropology and maintained close collaboration with the international academic community, particularly in Germany. Hoinathy's research initially focused on natural resources and their governance. He then researched various socio-political dynamics in Chad, including migration and conflict, religious diversity and radicalisation processes, religious-based violent extremism, and security governance. His priorities are the Lake Chad region and Central Africa. – hoinathy@googlemail.com

Jonathan Ngeh is a Post-doctoral Researcher and Principal Investigator at the Global South Studies Center (GSSC) of the University of Cologne, where he also serves as the speaker for the Co-producing Knowledge thematic group. He earned his PhD in sociology from the University of Umeå, Sweden, and specialises in transnational migration, focusing on the experiences of Africans in Northern Europe and the Gulf Cooperation Council (GCC) states. Ngeh has conducted extensive fieldwork in Cameroon, Sweden, and the United Arab Emirates. His research explores social inequalities, including labour exploitation, unequal power relations, and social exclusions. Before joining the GSSC in 2019, he taught Development Theory and Policy courses at the University of Bamenda, Cameroon. – jngeh@uni-koeln.de

Paul Nchoji Nkwi is professor of African anthropology at the University of Yaounde I, Cameroon. He studied anthropology and education at the University of Fribourg, Switzerland. In Cameroon, he taught anthropology at the University of Yaoundé, at the Catholic University of Central Africa, Yaoundé, and at the St. Thomas Aquinas Major Seminary Bambui, Bamenda. He was visiting professor at Harvard Medical School, Department of Social Medicine, University of South Carolina, Columbia as a Fulbright Scholar, at the University of Frankfurt, Frobenius Institut as a Alexander von Humboldt Fellow, Rhodes University, Grahamstown, South Africa,Visiting Fellow at the University of Leiden (1982),and Research Fellow at the University of Bergen, Centre for Social Policy, (Bergen, Norway). He is one of the founding members of the Pan African Association of Anthropologits (PAAA). – nkwi70@yahoo.com

Bakheit Mohammed Nur is a postdoctoral researcher at the Africa Multiple Cluster of Excellence, University of Bayreuth, Germany. He holds a Ph.D. in Anthropology from the Bayreuth International Graduate School of African Studies (BIGSAS), University of Bayreuth; and a B.A. in Sociology and Social Anthropology and an M.A. in Social Anthropology from the University of Khartoum, Sudan. In 2022, Nur was a Fellow of the Bayreuth Academy of Advanced African Studies, Africa Multiple Cluster of Excellence, University of Bayreuth. Between 2019 and 2021, Nur was a postdoctoral fellow at the Department of Anthropology, University of Bayreuth, where he worked on the project "Islam, Politics and Society: Islamist Education Policy and Epistemic Orientation in Sudan", funded by the Fritz Thyssen Foundation. Between 2018 and 2019, he was an assistant professor at the Institute of African and Asian Studies (IAAS), University of Khartoum. His research focuses on the anthropology of Islam, Islamic epistemology, education and knowledge, religion and conflict, Muslim religious specialists, automobility and technology. Nur's most recent publications include: *Religion, Politics, and Society: The Role of Political Islam in the Sudanese Revolution of December 2016 and its Aftermath*, Modern Africa: Politics, History and Society, 2023; *Politics of Epistemology in Postcolonial Africa: The Islamisation of Knowledge in the Sudan*, Politics, Religion & Ideology, 2022. – bakheit.m.nur@uni-bayreuth.de

Michaela Pelican works as a professor of social and cultural anthropology at the University of Cologne Cologne. She is the rector's representative for international affairs and the speaker of the Global South Studies Center Cologne (GSSC). She did her doctorate at the Martin-Luther University Halle-Wittenberg and was a junior researcher at the Max Planck Institute for Social Anthropology in Halle/Saale. Before coming to Cologne in 2012, she was a senior lecturer and researcher at the University of Zurich and a visiting fellow at the University of Kyoto. Her research focuses on migration, inequality, conflict, indigeneity and research methodology with a regional focus on West Africa, the Gulf States and China. – mpelican@uni-koeln.de

Djimet Seli is a research lecturer in the Anthropology Department at the University of N'Djamena and Administrative Director of the Centre de Recherches en Anthropologie et Sciences Humaines (CRASH). He conducts academic and action research on a number of themes, most notably on gender issues, social conflicts, social networks and relational dynamics, pastoralism and cohabitation. – djimet_selde@yahoo.fr

Antoine Socpa is affiliated with the Department of Anthropology at the University of Yaoundé. He holds a PhD in Social Anthropology from the University of Leiden (The Netherlands) and a Doctorate in Medical Anthropology from the University of Yaoundé (Cameroon). His research and publications are in the fields of cultural, political, social and medical anthropology. He is a senior researcher at the Center for Research and Training in Applied Social Sciences (CASS-RT), a consulting laboratory based in Cameroon. He is the Secretary General of the Pan-African Association of Anthropologists (PAAA). He was visiting professor at ASAFAS, Kyoto University in Japan (2016) and visiting researcher (2023) thanks a grant from Canon Foundation in Europe. – asocpa@yahoo.com

Karim Zafer works as a project manager and is responsible for the coordination of academic collaboration with the MENA-Region at the Division for International Affairs at the University of Cologne. He is also a lecturer at the Department of Social and Cultural Anthropology at the University of Cologne, from which he received his PhD in 2024 and Master's degree in 2016. Before migrating to Germany in 2013, he studied and worked in the fields of tourism and Egyptology in Egypt. His research interests include (forced) migration, youth, decolonial research methodology, and the MENA region. – karimzafer@icloud.com